The Digital Satellite TV Handbook

The Digital Satellite TV Handbook

Mark E. Long

Newnes

Boston • Oxford • Auckland • Johannesburg • Melbourne • New Delhi

Newnes is an imprint of Butterworth-Heinemann.

 A member of the Reed Elsevier group

∞ Recognizing the importance of preserving what has been written, Butterworth-Heinemann prints its books on acid-free paper whenever possible.

 Butterworth-Heinemann supports the efforts of American Forests and the Global ReLeaf program in its campaign for the betterment of trees, forests, and our environment.

Library of Congress Cataloging-in-Publication Data

Long, Mark.
    The digital satellite TV handbook / Mark E. Long.
        p.   cm.
    Includes index.
    ISBN 0-7506-7171-8 (alk. paper)
        1. Direct broadcast satellite television.   2. Digital television.
    I. Title.
    TK6677.L64   1999
    621.388′53—dc21   99-13017
                                                        CIP

British Library Cataloguing-in-Publication Data
A catalogue record for this book is available from the British Library.

The publisher offers special discounts on bulk orders of this book.
For information, please contact:

Manager of Special Sales
Butterworth-Heinemann
225 Wildwood Avenue
Woburn, MA 01801-2041
Tel: 781-904-2500
Fax: 781-904-2620

For information on all Newnes Press publications available, contact our World Wide Web home page at: http://www.newnespress.com

10 9 8 7 6 5 4 3 2 1

Printed in the United States of America

Contents

Foreword

On June 25, 1967, for two hours 26 nations of the world were joined together by an invisible electromagnetic grid utilizing four satellites. The London-based production, in glorious black and white, was the first-ever use of satellites to simultaneously interconnect remote corners of the world to a single program event. The program, appropriately entitled "Our World," included the Beatles debuting the song "All You Need Is Love" to an audience estimated at more than 600 million.

During the course of the telecast, live feeds were interconnected through a pair of early design Intelsats, an American experimental satellite (ATS-1), and a Russian Molniya class bird. The *New York Times* would write about the ground-breaking telecast, "'Our World' was a compelling reaffirmation of the potential of the home screen to unify the peoples of the world."

Less than three decades later, or approximately the period of one generation of mankind, more than 30 million homes in the world are equipped with their own satellite dishes. The early Intelsat, ATS, and Molniya satellites were capable of relaying one (or at most, two) simultaneous TV programs; each satellite of the current generation easily can deliver as many as 200 program channels to dish antennas less than one-thirtieth of the size required for reception of the original "Our World" telecast.

Well before the turn of the century, virtually any location in Asia or the Pacific will have direct access to hundreds of channels of TV, high-speed Internet links, and thousands of radio program channels. It is not an exaggeration to suggest that satellites are redesigning the very fabric of life by creating full-time universal access to "our world."

All of this technology creates virtually unlimited opportunities for new business enterprise and personal development. You are holding in your hand a key that will unlock for you, your family, and your business the "secrets" of the 21st century "Information Revolution." There has never been a point in the history of the world when so much opportunity has presented itself to mankind. Use what you learn here wisely and your life will forever be changed.

Robert B. Cooper
Managing Director, SPACE Pacific Trade Association,
Mangonui, New Zealand

About the Author

Mark E. Long is the author of numerous technical books and reference manuals, as well as hundreds of magazine articles on radio, television, and satellite telecommunication technologies. A graduate of the School of Communications at the University of Central Florida (UCF), Mark was the transmitter manager for radio station WUTZ-FM during the late 1970s. From 1979 to 1981, his service as a telecommunication consultant for Greenpeace International included several assignments as the shipboard radio operator on the Greenpeace flagship *Rainbow Warrior*. As a telecommunications supervisor at Solar Electronics International, he developed two-way, solar-powered telecommunications systems for rural applications, designed satellite TV receiving systems, and prototyped radiation-detection equipment marketed under the Radiation Alert brand name.

Over the years, Mark has served as the technical editor of *Satellite Orbit* magazine; the U.S. Bureau Chief of *Satellite World* magazine; satellite columnist at *Popular Communications* magazine; and contributing editor to *Satellite Direct, TVRO Dealer, Orbit Video,* and *Satellite Retailer* magazines. He also has been a frequent contributor to *Cable & Satellite Asia, Middle East Satellites Today, SATFacts,* and *Tele-satellit* magazines. More than 350 of his magazine and newspaper articles have been printed to date.

Mark provides a wide variety of consulting services for the worldwide satellite telecommunications industry. During the early 1980s, he worked closely with numerous educational institutions in the United States to adapt the use of foreign-language satellite television broadcasts for language learning applications. Among his consulting clients on this project were the U.S. Naval Academy; the Universities of Maryland, Pennsylvania, and Virginia; and the SCOLA foreign-language distance learning program developed at Creighton University.

His early interest in international satellite communications also brought him to the attention of various Latin American companies that hired him as a consultant to assist in the development of DTH businesses in Latin America. He was the first person to install a DTH satellite TV system in Bolivia back in 1983 and provided on-site technical training and guidance for companies in Argentina, Bolivia, and Mexico in 1983 and 1984. In 1982, Mark also was the first person to publicly display the potential of home satellite TV technology to British consumers attending an electronics show at the Wembley Exhibition Hall in London.

During the past two decades, Mark has personally conducted satellite TV technology workshops and training sessions for more than 2,000 individuals at trade conferences, universities, and private corporations located on three continents. Past training seminar clients include the trade associations CASBAA, the Satellite Broadcasting Communications Association (SBCA), and SPACE Pacific; the educational associations SCOLA and the U.S. Distance Learning Association; and private corporations such as IIR Conferences Pte Ltd., Measat Broadcast Network Systems, and STAR TV. Mark continues to teach individual students through a series of correspondence courses managed by the SPACE Pacific cable and satellite trade association. He may be contacted at his Internet site on the World Wide Web at http://www.mlesat.com or by e-mail at mlesat@mlesat.com.

Introduction

The Digital Satellite TV Handbook represents a totally new approach to educating satellite TV enthusiasts and professionals about satellite communications technology. This new publication presents the important technical concepts in a highly visual way, which is appropriate given that TV itself is a visual medium. If a single picture is indeed worth a thousand words, then the graphic information presented herein far exceeds the informational content of the text itself.

Redundancy—duplication of components within the communications payload—is a common feature of all telecommunication satellites. In the event that a primary module fails, a replacement can be activated to maintain continuity of operations. Like the satellites themselves, this book also incorporates a certain amount of redundancy, for instance, certain key concepts are presented more than once. As a teacher I have discovered that not everyone will understand a key concept the first time that it is presented. Whenever key concepts are revisited in this publication, they are put into the context of the subject matter of the chapter in which they appear.

The CD-ROM that accompanies this handbook allows readers to access larger, full-color versions of selected graphics from this publication for their personal use. All of the computer graphic files on these disks are themselves the result of digital compression. Moreover, the CD-ROM also contains numerous hyperlinks to additional reference materials that may be accessed from the Internet by those readers equipped with web-browser software and a connection to the World Wide Web.

A one-hour Satellite Installations videotape also is available in either the PAL or NTSC video format. This videotape is part of a self-study course that I have developed for the SPACE Pacific cable and satellite trade association serving Australia, New Zealand, and the South Pacific. The videotape features graphic animations and on-camera demonstrations relating to various concepts presented in this handbook.

With the exception of Chapter 3, each chapter ends with a series of "Quick Check Exercises" that provides readers with the opportunity to confirm their understanding of the key concepts presented. An answer key to each chapter's "Quick Check Exercises" is provided in the Appendix. For those readers who are using this textbook as part of a trade association or private industry certification course, these exercises are the perfect way to prepare for the installer and technician certification exams.

Some of the information presented in this handbook, including digital DTH service descriptions and satellite transponder data, is extremely time-sensitive. Internet links are provided throughout so that readers can access the very latest information on satellites and digital satellite TV services. Updates to the contents of this book also are posted periodically at my Internet site at http://www.mlesat.com.

The creation of this handbook and its supplemental course materials would not have been possible without the kind assistance of numerous friends and associates around the world. Thanks to David and Ellen Shelburne for producing the one-hour Satellite Installations videotape that supplements this publication and for making suggestions concerning the presentation graphics that appear in both the book and videotape.

Over the years, my knowledge of satellite TV technology has expanded due in great measure to the kind assistance of several satellite industry experts and pioneers. My heartfelt thanks to Stephen J. Birkill, Arthur C. Clarke, Bob Cooper, Brett Graham, H. Taylor Howard, and Jim Roberts for their input and support.

Perhaps the single most important contribution to this book has been the many questions that I have had to answer at workshops and training seminars during the past two years. A wise man once said that the best way to learn is to teach. I should like to thank the hundreds of students with whom I have had the honor of interacting. This book is dedicated to them.

Chapter 1

Satellite Frequencies and Orbital Assignments

Whenever we listen to radio broadcasts, watch TV, use a cellular phone or talk on a two-way radio, invisible waves of electromagnetic energy are bringing us messages from distant locations. These invisible waves continually bombard us as we walk down the street, play sports or putter about the garden. We only become aware of the electromagnetic soup that surrounds us if we have the right antenna and receiver for tuning in to these signals.

FREQUENCY VERSUS WAVELENGTH

At the beginning of the twentieth century, Marconi discovered that it was possible to modulate invisible waves of electromagnetic energy so that they could carry messages as they radiate through space at the speed of light. For the first time ever, humanity was able to communicate over vast distances almost instantaneously. By the late 1920s, millions of people around the world were tuning in to AM radio stations: the world's first electronically generated virtual realities. Back then, these transmissions were simply called radio waves. As time went on, however, it became apparent that waves of electromagnetic energy could be used to transmit all sorts of information, including TV pictures.

In many respects, electromagnetic waves are similar to waves on the ocean. In one complete cycle of a wave, water rises from sea level until it swells to reach the crest of the wave, then plummets downward into the wave trough before rising again to sea level. A communication signal is the electromagnetic equivalent of a message in a bottle: it rides a never-ending succession of waves before arriving at its final destination.

The frequency of any communications signal is the number of cycles per second at which the radio wave vibrates or "cycles." Electromagnetic waves cycle at phenomenal rates: one thousand cycles per second is called a kilohertz (kHz), one million cycles per second

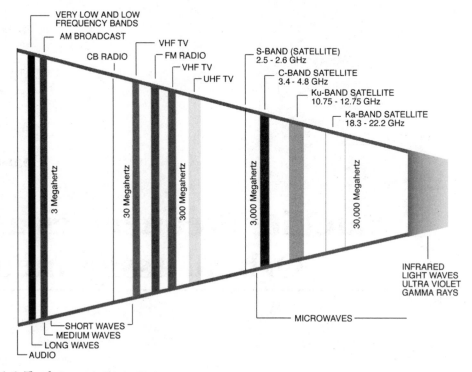

Figure 1–1 The electromagnetic spectrum.

a megahertz (MHz), and one billion cycles per second a gigahertz (GHz). Today we refer to the continuum of frequencies used to propagate communications signals—100 kHz to 100 GHz and beyond—as the electromagnetic spectrum (Figure 1–1).

The distance that each wave travels during a single cycle is called its wavelength. There is an inverse relationship between frequency and wavelength: the higher the frequency, the shorter the wavelength (Figures 1–2 and 1–3).

Each subset or band of frequencies within the electromagnetic spectrum has unique properties that are the result of changes in wavelength. For example, medium wave signals (500 kHz to about 3 MHz) radiate along the surface of the Earth over hundreds of miles, which is perfect for relaying AM radio stations throughout a region.

International radio stations use the short wave bands (3–30 MHz) to span distances of thousands of miles. The ionosphere—upper layers of the Earth's atmosphere that are electrically charged by the Sun—reflects these short waves back down to Earth, much as a

The actual distance that an electromagnetic radio wave travels during a complete cycle from 0 to 360 degrees is called the wavelength

The frequency of the radio wave is expressed in cycles per second (cps), or "hertz."

1 cps = 1 hertz
1,000 cps = 1 kilohertz
1 million cps = 1 megahertz
1 billion cps = 1 gigahertz

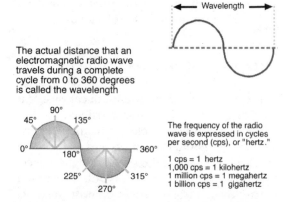

Figure 1–2 Frequency and wavelength have an inverse relationship: the higher the frequency, the shorter the wavelength.

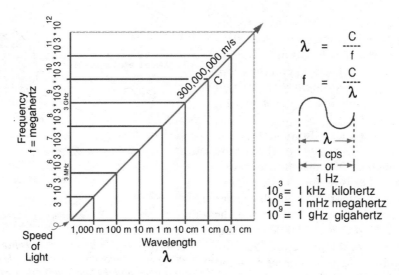

$$\lambda = \frac{C}{f}$$

$$f = \frac{C}{\lambda}$$

|← λ →|

1 cps
|← or →|
1 Hz

10^3 = 1 kHz kilohertz
10^6 = 1 mHz megahertz
10^9 = 1 gHz gigahertz

Figure 1–3 *This relationship between frequency and wavelength can be calculated using the formulas presented here.*

mirror or any other shiny metal object can reflect beams of light (Figure 1–4).

TV and FM radio broadcasters use the very high frequency (VHF) and ultra high frequency (UHF) bands, located from 30 MHz to 300 MHz and 300 MHz to 900 MHz, because these signals only cover short distances; they can't travel very far along the Earth's surface or skip off the ionosphere. The advantage to using these frequency bands for local communications is that dozens of TV and FM radio stations can use identical frequencies within any given country or region without causing interference to one another.

SATELLITE ORBITAL ASSIGNMENTS

Mathematician and science fiction author Arthur C. Clarke first proposed the use of satellites for TV and radio broadcasting in a pioneering article that appeared in the October 1945 issue of *Wireless World* magazine. A central part of what Arthur now calls his "modest proposal" was the use of a unique orbit some 22,300 miles above the Earth's equator. "A body in such an orbit," he wrote, "would revolve with the earth and would thus remain stationary above the same spot on the planet called the sub satellite point. It would remain fixed in the sky of a whole hemisphere." This unique orbit is now known as the geostationary arc.

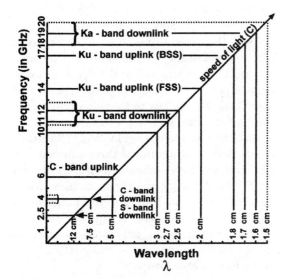

Figure 1–4 *The satellite frequency bands have wavelengths that are so short that these signals are called microwaves.*

The importance of the geostationary arc cannot be overstated. Satellites in geostationary orbit remain at a specific point in the sky relative to any receiving location and therefore can be received by a fixed satellite antenna 24 hours a day without any need to ever repoint the dish.

Meridians, imaginary lines circling the Earth from pole to pole, cross over each of the equator's 360 degrees. The distance from one meridian to any other is defined in degrees of longitude. The prime meridian, which crosses through London, England, is referred to as zero degrees, and the two 180-degree segments to either side of the prime meridian are assigned ascending values of degrees east and west longitude. East and west meet again at the International Date Line that runs through the Pacific Ocean.

References to any satellite's orbital location, as well as to the intervals between adjacent satellites, are made in degrees of longitude. Keep in mind, however, that the geostationary orbit is a circle and the reference point for the calculation of degrees longitude is the Earth's center. Since we all live on the surface of the Earth, the apparent spacing between two satellites will be greater than the actual spacing in degrees of longitude (Figure 1–5).

The precise amount of variance between actual and apparent spacing between two satellites is a function of site latitude and the difference between site longitude and satellite longitude. For the purpose of calculating the antenna performance charts that appear later in this chapter, I have assumed an apparent spacing of 3.4 degrees between satellites that are separated by 3 degrees in longitude.

ORBITAL SPACING REQUIREMENTS

Two degrees of orbital spacing between adjacent satellites serving the same geographical area is required to prevent interference between satellite communication systems with

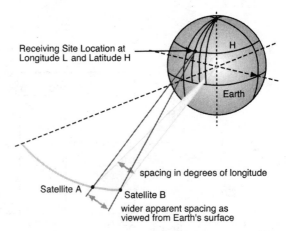

Figure 1–5 The apparent spacing between satellites is a function of the receiving site's longitude and latitude coordinates and the longitude of the two satellites under observation.

overlapping coverage areas. In the case of adjacent digital DTH satellites, the minimum spacing requirements may be even greater. This is because small-aperture receiving antennas (less than 1.2 m in diameter) have a beam width that can intercept signals from adjacent satellites as well as the intended satellite.

The International Telecommunication Union (ITU) of the United Nations is the global agency responsible for the allocation of satellite orbital assignments and telecommunication frequencies on a global basis. A perusal of the ITU's 71-page List of Geostationary Space Stations on the Internet (http://www.itu.int/itudoc/itu-r/space/snl/sect1_b_21798.html) shows that a minimum of three national administrations have submitted ITU filings for most orbital locations. These include countries such as Cuba, Iraq, and the Kingdom of Tonga, or even small territories such as Cyprus, which have generated dozens of "paper satellites" with little prospect of flying in the foreseeable future.

As a result, many administrations now file for more orbital slots than they actually need in the expectation that any extra locations can be used as bargaining chips in negotiations with other satellite operators.

Until recently, the ITU had no authority to judge the viability of individual satellite filings or act as a referee whenever orbital conflicts arise. The ITU's 1997 World Radiocommunication Conference (WRC-97), however, adopted a resolution targeting paper satellites that needlessly hoard scarce orbital resources as well as lengthen and complicate the process of registering and coordinating new satellite systems. Under Resolution 18, national administrations filing for new orbital slots and frequencies must demonstrate due diligence by disclosing the identity of the satellite manufacturer, satellite operator, and launch vehicle provider, as well as disclosing the satellite's contractual launch date.

Satellite operators also are implementing technical solutions to alleviate orbital congestion. These include the colocation of multiple satellites at the same orbital location to create a single satellite "constellation" and the use of new "extended C-band" expansion bands.

A prime example of a Ku-band satellite system using colocation is the Astra satellite constellation at 19.2 degrees east longitude (Figure 1–6). A total of six Astra spacecraft, each operating within a distinct segment of the

10.7–12.75 GHz satellite frequency band, are colocated at a single orbital assignment. One major advantage of satellite collocation: a single fixed antenna can receive hundreds of digital TV and radio services from the satellite constellation without having to change its alignment.

The Thaicom satellites colocated at 78.5 degrees east longitude collectively offer 22 standard C-band (3.7 GHz to 4.2 GHz) transponders plus 12 extended C-band (3.4 GHz to 3.7 GHz) transponders. (See Figure 1–7.) Both Thaicom satellites also carry Ku-band transponders downlinking signals between 12.25 GHz and 12.75 GHz. The Indian Satellite Research Organization (ISRO) also has the Insat 2B and 2C satellites colocated at 93.5 degrees east longitude. Each satellite contributes 12 standard C-band transponders as well as six extended C-band transponders that downlink in the 4.5 GHz to 4.8 GHz expansion band.

Advanced spacecraft designs and a new generation of powerful launch vehicles are making it possible for satellites to carry a greater number of transponders than ever before. Leading manufacturers are switching to ion propulsion systems, which dramatically reduce the amount of station-keeping fuel that

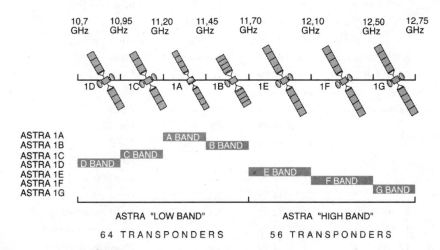

Figure 1–6 *The Astra satellite constellation at 19.2 degrees east longitude.*

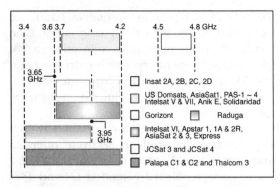

Figure 1–7 *C-band satellites downlink in different segments of the 3.4–4.8 GHz spectrum. Extended C-band transponders operate below 3.7 GHz or above 4.2 GHz.*

each satellite must carry into orbit. Satellite designers can therefore increase the weight of each satellite's communications payload to accommodate additional transponders and multiple spot beam antennas.

In 1998, Telesat Canada ordered the construction of an Anik replacement satellite that will carry 84 active transponders: 48 operating in the Ku-band satellite frequency range and 36 operating in the C-band spectrum. Expect

operators in other regions to launch similar replacement satellites in the early years of the new millennium to more efficiently use the orbital locations and frequency spectrums already under their control.

SATELLITE FREQUENCY ASSIGNMENTS

The ITU has set aside spectrum in the super high frequency (SHF) bands located between 2.5 GHz and 22 GHz for geostationary satellite transmissions (see Table 1–1). At these frequencies, the wavelength of each cycle is so short that the signals are called microwaves. These microwaves have many characteristics of visible light—they travel directly along the line of sight from the satellite to its primary coverage area and are not impeded by the Earth's ionosphere.

The scientists who developed the first microwave radar systems during World War II assigned a letter designation to each microwave frequency band. For example, the 800 MHz to 2 GHz frequency range was called the

Table1–1 World Satellite Frequency Assignments

Uplink	Downlink	Band	Service Type
5.855–6.055	2.535–2.655	S	Broadcast
5.725–5.925	3.400–3.700	Extended C	Fixed
5.925–6.425	3.700–4.200	C	Fixed
6.425–7.075	4.500–4.800	Extended C	Fixed
7.900–8.400	7.250–7.750	X	Military
12.75–13.25	10.70–10.95	Ku	Fixed
14.00–14.50	10.95–11.20	Ku	Fixed
	11.20–11.45	Ku	Fixed
	11.45–11.70	Ku	Fixed
	11.70–12.20	Ku	Fixed (Americas)
	11.70–12.25	Ku	Fixed (Asia)
	12.50–12.75	Ku	Fixed
17.30–17.80	12.25–12.75	Ku	Fixed (Asia/Pacific)
17.30–17.80	12.20–12.70	Ku	Broadcast (Americas)
17.30–18.10	11.70–12.50	Ku	Broadcast (Europe)

"L" band. The 2 GHz to 3 GHz spectrum is called the "S" band. The 3 GHz to 6 GHz band is called the "C" band. The 7 GHz to 9 GHz frequency band is called the "X" band. The 10 GHz to 17 GHz spectrum is called the "Ku" band (Figure 1–8). The 18 GHz to 22 GHz frequencies are known as the "Ka" band. At the dawn of the Satellite Age during the mid-1960s, microwave engineers decided to carry forward the existing radar terminology and apply it to the communications satellite bands as well.

The world's first commercial satellite systems used the "C" band frequency range of 3.7 GHz to 4.2 GHz. By the late 1960s, many telephone companies around the world had numerous terrestrial microwave relay stations that also operated within the same frequency range. The amount of power that any C-band satellite could transmit therefore had to be limited to a level that would not cause interference to terrestrial microwave links.

The first commercial "Ku" band satellites made their appearance in the late 1970s and early 1980s. Relatively few terrestrial communications networks were assigned to use this frequency band; Ku-band satellites therefore could transmit higher-powered signals than their C-band counterparts without causing interference problems down on the ground.

SATELLITE TRANSPONDERS

A geostationary telecommunication satellite can be compared to a communications tower that is 22,300 miles tall. From its vast height, a single satellite can transmit a signal that potentially can cover 42.4 percent of the Earth's total surface.

Every satellite is a repeater of a communications signal. Satellite channels, called transponders, consist of an on-board receiver, which processes the signal that the originating earth station transmits or "uplinks" to the spacecraft using one set of frequencies, and a transmitter, which repeats or "downlinks" the informational content of the original signal using a second set of frequencies (Figures 1–9 and 1–10). A complete C-band satellite transmission, for example, actually consists of a pair of frequencies: an earth station uplinks the signal to the satellite using the 6 GHz frequency band and the satellite downlinks the signal using the 4 GHz frequency band. Each satellite transponder has a finite amount of frequency spectrum or bandwidth that is used to relay one or more telecommunications signals.

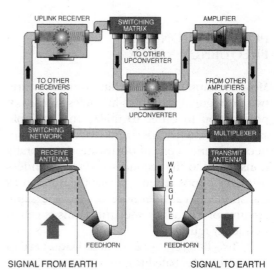

Figure 1–9 A satellite transponder is the combination of an uplink receiver and a downlink transmitter that relays or repeats one or more communication signals.

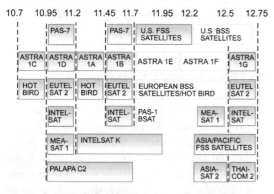

Figure 1–8 FSS and BSS satellites are assigned to different segments within the Ku-band frequency range, depending on the ITU region that the satellite is intended to serve.

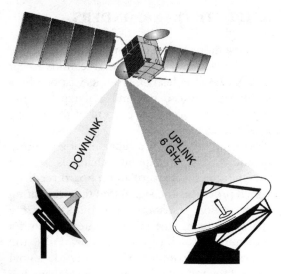

Figure 1–10 The transponder uses a paired set of frequencies: one for receiving the uplink signal and the other for transmitting the downlink signal.

Satellite transponder bandwidths may range from 24 MHz to 108 MHz. The maximum permissible data rate through any satellite transponder is a direct function of the bandwidth that is available.

POLARIZATION AND FREQUENCY REUSE

Most communications satellites transmit using two orthogonal (i.e., at right angles) senses of polarization in order to utilize the available satellite frequency spectrum twice. Transponders with one sense of polarization are totally transparent to the second set of transponders using the opposite sense. Twice the number of transponders can therefore occupy the same amount of frequency spectrum. This is called frequency reuse.

Two different types of polarization are in use worldwide. Satellites using linear polarization orient the signal wave front along either a horizontal or vertical plane from the perspective of the satellite platform, while satellites

using circular polarization rotate the signal wave front in either a clockwise or counterclockwise direction from the perspective of the satellite platform.

SATELLITE POWER LEVELS

Satellite signal strength—called the effective isotropic radiated power (EIRP)—is expressed in decibels referenced to one watt of power (dBW). An increase of 3 dBW represents a doubling of power; 10 dBW represents a tenfold increase; 20 dBW a one-hundred-fold increase; and so on.

C-band satellites typically transmit signal levels ranging from 31 dBW to 40 dBW. The strongest signals fall within the center of each satellite's coverage beam, with signal intensity decreasing outward from there. Depending on the location of the receiving site within the satellite's primary coverage beam or footprint, the antenna apertures required to receive crystal-clear TV pictures typically range from 1.8 m to 3.7 m in diameter.

Ku-band satellites transmit nominal signal levels ranging from 47 dBW to 56 dBW—a 16-dBW increase in power over what most C-band satellites can deliver. Receiving antennas as small as 30 cm in diameter can therefore be used to receive Ku-band satellite signals. This significant reduction in antenna size lowers the cost of the receiving equipment and simplifies the system installation requirements.

ANTENNA BEAM WIDTH

The output power of a satellite transponder is not the sole factor that determines the minimum antenna diameter that may be required to receive any satellite transmission. The beam width of the receiving antenna, that is, the narrow corridor through which the dish looks up at the sky, also defines the minimum antenna diameter that can function within a given satellite frequency band (Figure 1–11). There is a

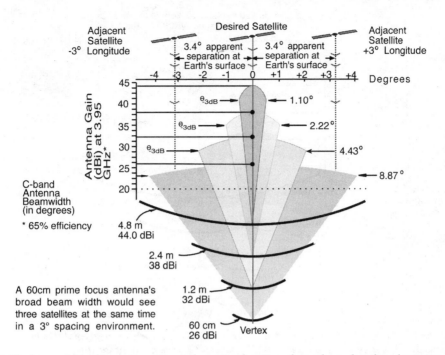

Figure 1–11 *The beam width of an antenna's main beam is a function of signal wavelength and antenna diameter.*

direct relationship between signal wavelength and antenna beam width: the shorter the wavelength, the narrower the beam width.

A 60 cm C-band antenna could potentially receive three satellites within its main beam if the satellites are separated by 3 degrees in longitude. In the United States, where 2-degree spacing between satellites is now a regulatory requirement, a 60 cm C-band antenna could potentially have signals from four satellites falling within its main beam. When receiving Ku-band satellite signals, however, 30–60 cm dishes become possible because of the dramatic reduction in antenna beam width that takes place when we use the higher satellite frequency bands (Figure 1–12).

Although the International Telecommunication Union has assigned S-band frequency spectrum for direct-to-home TV transmissions, few organizations have so far elected to use this frequency range for satellite TV broadcasting. One limiting factor has been the limited bandwidth available: just 100 MHz of spectrum from 2.5 GHz to 2.6 GHz. The ISRO and the Arabsat organization have included S-band transponders on a few of their C-band satellites to make it cost effective to launch S-band payloads into geostationary orbit.

In 1997, Indonesia launched Cakrawartha-1, the world's first dedicated S-band satellite, to an orbital assignment of 107.7 degrees east longitude. Digital video compression has made it technically and economically feasible for Indostar to broadcast a multichannel TV package to subscribers in Indonesia.

Indostar uses antennas ranging from 70 cm to 1m in diameter for DTH reception. DTH systems such as Indostar, however, will only work if there are no other adjacent satellites using the same frequency spectrum (Figure 1–13). Even a 1.2-m antenna would have problems in a 3-degree spacing environment. S-band do, however, experience minimal rain fade problems. This is a significant advantage for a satellite broadcast system serving the world's highest rain rate region.

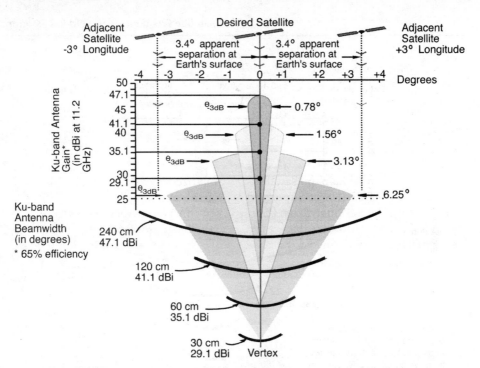

Figure 1–12 *Ku-band satellites can provide digital DTH services into 60 cm or even 30 cm dishes because the beam width for these antennas is narrow enough to reject signals coming from adjacent satellites.*

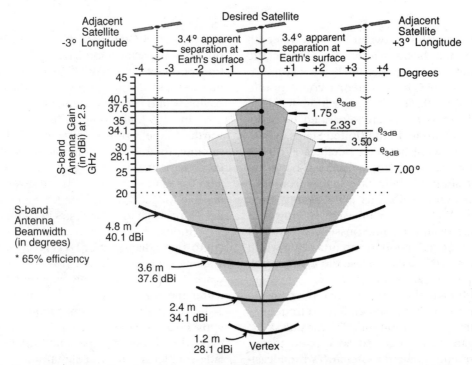

Figure 1–13 *S-band antennas have broad beam widths and therefore can be used to receive digital DTH services only when there are no other S-band satellites operating at adjacent orbital locations.*

Sixteen years ago, the number of commercial Ku-band communications satellites orbiting the Earth could be counted off using the fingers of one hand. Today there are more than 75 Ku-band satellites in operation worldwide. Within the past few years, satellite operators have begun exploring the brave new world at 20 GHz. Only a few Ka-band satellites are currently in orbit: ACTS (USA); Superbird and N-STAR (Japan); DFS Kopernikus (Germany); and Italsat (Italy). However, expect the use of this higher frequency band to increase dramatically during the first decade of the twenty-first century (Figure 1–14).

RAIN ATTENUATION

There is one major drawback to satellites downlinking signals at frequencies greater than 10 GHz: the length of the microwaves is so short that rain, snow, or even rain-fill clouds passing overhead can reduce the intensity of the incoming signals (Figure 1–15). At these higher frequencies, the lengths of the falling rain droplets are close to a resonant submultiple of the signal's wavelength; the droplets therefore are able to absorb and depolarize the microwaves passing through the Earth's atmosphere.

In places such as Southeast Asia or the Caribbean, torrential downpours can lower the level of the incoming Ku-band satellite signal by 20 dB or more; this will severely degrade the quality of the signals or even interrupt reception entirely. The duration of rain outages, however, is usually very short and typically occurs in the afternoons or early evenings rather than during the prime-time viewing hours. For most Ku-band satellite TV viewers, these service interruptions will only amount to a few hours of viewing time over the course of any given year.

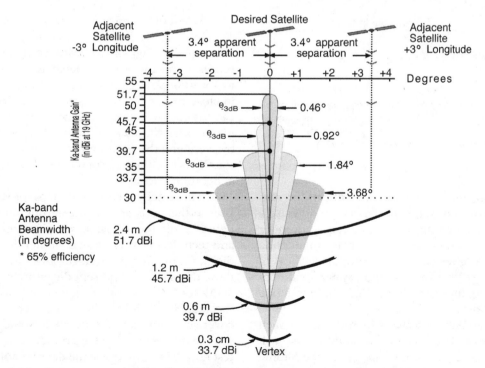

Figure 1–14 *Antennas receiving Ka-band satellite signals have extremely narrow beam widths.*

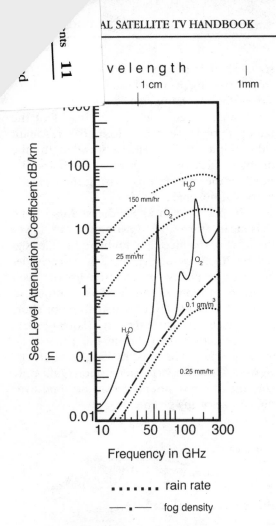

Figure 1–15 Rain attenuation in decibels per kilometer for frequencies above 10 GHz. For certain frequencies, even water and oxygen molecules in the atmosphere also can attenuate communication signals.

To help counteract the effects of rain fade, Ku-band system designers typically use a larger antenna than would be required under clear sky conditions. This increase in antenna aperture gives the system several decibels of margin so that the receiving system will continue to function during light to moderate rainstorms.

Satellite TV viewers in arid regions such as central Australia or the Middle East will rarely experience rain outages. In the Middle East, however, satellite dish owners may expe-

rience outages caused by intense sandstorms. The presence of any atmospheric particulate—even sand—can have an adverse effect on satellite TV reception.

SATELLITE FOOTPRINTS

All geostationary satellite operators generate for their customers one or more maps—called footprints—that display the intended coverage area for one or more beams generated by the spacecraft. The contour lines on any coverage map show the signal level at specific geographical locations. The satellite map's transmit contours provide the value of effective isotropic radiated power (EIRP) or power flux density (PFD), while the spacecraft's receive footprint map shows the signal level in G/T or saturated flux density (SFD). The strongest signal levels usually will be found toward the center of the footprint; lower signal levels will be found outward from there until the beam edge of beam pattern is reached.

When planning any receiving system, the satellite installer and technician may refer to transmit coverage maps, with each contour representing the EIRP in decibels referenced to 1 watt of power (dBW) for locations located in close proximity to the contour line. For locations that lie between two contour lines, the map reader will need to interpolate the value by averaging the values given for the two contour lines to either side of the installation site's geographical location.

The satellite coverage maps presented later in this book are a good starting point for determining the signal levels that installers and technicians will likely encounter down on the ground. For several reasons, however, these maps may not present the entire picture. Many footprint maps are generated prior to the launch of the satellites themselves. The coverage contours are derived from range tests that are conducted at an indoor antenna test range. In some cases, the data are adjusted following in-orbit testing, where precise signal measurements are made at a number of earth stations located throughout the satellite coverage beam. Whenever possible, technicians and installers are advised to supplement the information that these maps provide with data from recent industry publications as well as input from other satellite professionals working in the same general area.

On occasion, unexpected signal levels will be encountered at locations that lie outside of the footprint map's coverage contours. Out-of-beam coverage may permit the reception of signals at locations that are not within the satellite operator's intended zone of coverage. For example, the Astra satellites collocated at 19.2 degrees east longitude have beam patterns that were supposed to restrict coverage to Europe, North Africa, and portions of the Middle East. Satellite earth stations located throughout southern Africa, however, also can receive signals from these satellites. This phenomenon is known as signal "spillover."

The contour lines to be found on all satellite coverage maps also represent saturated values—that is, the values that would be generated if the satellite transponder operated at full power. In many cases, however, the satellite uplink engineers will back off the uplink power from saturation. For example, this may be done when multiple carriers are transmitted through the same transponder. Engineers may back off the power to prevent the generation of unwanted intermodulation distortion products between the various carriers. In the case of those satellites that use high-power traveling wave tube amplifiers, the maximum power output also may degrade over the lifetime of each tube.

Satellite coverage beams can be classified as one of five distinct types. The broadest beam—called the global beam—provides coverage of the 42.4 percent of the Earth's surface that any satellite can illuminate from its assigned orbital position. INTELSAT, Intersputnik, and other international satellite service providers use the global beam to bridge the world's major ocean regions as well as to provide telecommunications to islands in the Atlantic, Pacific, and Indian Ocean regions (Figure 1–16). The primary disadvantage of the global beam is that it spreads the available signal over a vast area, thereby lowering the signal level that can be delivered to any one location. Some geostationary satellites, such as Thaicom 3 and Apstar 2R, use what is known as a semiglobal beam pattern (Figure 1–17). The semiglobal beam concentrates the transmit signal over the Earth's land masses, while minimizing signal delivery to sparsely populated islands in the Indian Ocean region.

The hemispheric beam restricts satellite coverage to an area that represents approximately 20 percent of the Earth's surface (Figure 1–18). Two separate hemispheric beams aboard the same satellite can reuse the same frequencies and polarization senses within their respective geographically isolated coverage areas. This fourfold frequency reuse is called "spatial isolation."

Most INTELSAT satellites simultaneously transmit through spatially isolated east and west hemispheric beams that illuminate the

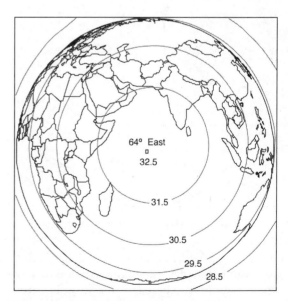

Figure 1–16 *INTELSAT VIII global beam (IOR).*

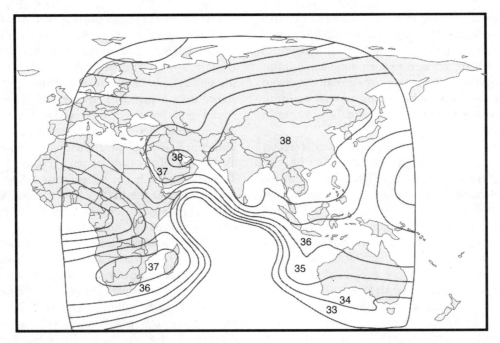

Figure 1–17 *Apstar 2R C-band Semi-global downlink beam (EIRP in dBW).*

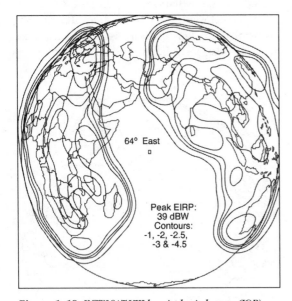

Figure 1–18 *INTELSAT VIII hemispheric beams (IOR).*

covers locations that are above the Earth's equator. The hemispheric beam obtains a 3-dB improvement over the global beam by concentrating the available power into an area that is less than half of the area covered by the global beam.

Additional improvements in signal concentration can be obtained by restricting the beam coverage area to even smaller dimensions. The newer INTELSAT satellites, for example, carry four spatially isolated beams that illuminate the northeast, northwest, southeast, and southwest quadrants of the Earth's surface that are visible from each satellite's orbital location. These so-called zone beams restrict coverage to areas that represent less than 10 percent of the Earth's surface (Figure 1–19). Highly focused coverage patterns, called spot beams, also may be used to boost signal levels within a particular region, or even a single country. INTELSAT satellites carry steerable Ku-band spot beams that are elliptical in shape (Figure 1–20). These beams illuminate the Earth's surface in a way that is

eastern and western halves of the Earth that are visible from each satellite's location in geostationary orbit, whereas some Russian satellites use a northern hemispheric beam that

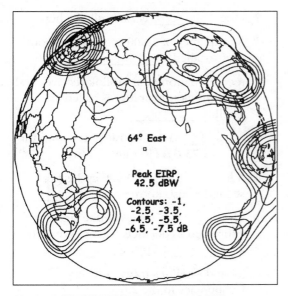

Figure 1–19 INTELSAT VIII Zone Beams (IOR).

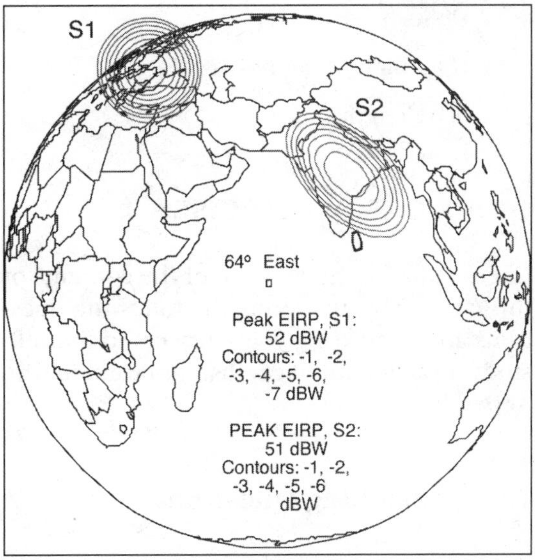

Figure 1–20 INTELSAT VIII Ku-band Spot Beams (IOR).

similar to the beam produced by a simple flashlight. Other communication satellites used shaped beams that have unique coverage patterns that maximize signal delivery over highly populated areas, while minimizing the illumination of underpopulated ocean or land-mass areas.

KEY TECHNICAL TERMS

The following key technical terms were presented in this chapter. If you do not know the meaning of any term presented below, refer back to the place in this chapter where it was presented or refer to the Glossary before taking the quick check exercises that appear below.

Apparent spacing

Band

Bandwidth

Beam width, antenna

Colocation

Cycles per second

Downlink

Effective isotropic radiated power (EIRP)

Electromagnetic spectrum

Frequency

Frequency reuse

Geostationary orbit

Gigahertz

Kilohertz

Medium wave

Megahertz

Microwaves

Polarization

Rain attenuation

Short wave

Subsatellite point

Transponder

Very high frequency (VHF)

Uplink

Ultra high frequency (UHF)

Wavelength

QUICK CHECK EXERCISES

Check your comprehension of the contents of this chapter by answering the following questions and comparing your answers to the self-study examination key that appears in the Appendix.

Part I: Matching Questions

Put the appropriate letter designation—a, b, c, d, etc.—for each term in the blank before the matching description.

a. kilohertz

b. megahertz

c. gigahertz

d. EIRP

e. Ku-band

f. C-band

g. watt

h. downlinks

i. beam width

j. apparent spacing

k. frequency

l. cycle per second

m. wavelength

n. transponder

o. subsatellite point

p. none of the above

1. The spot on the Earth's equator over which a geostationary satellite is positioned is called the _____.

2. One _____ is also called a hertz.

3. A frequency of 1,000 cycles per second is also called a _____.

4. A combination of an uplink receiver and a downlink transmitter is called a satellite_____.

5. _____satellites transmit within the 10.– 12.75 GHz frequency range.

6. A frequency of 1,000,000 cycles per second is also known as a _____.

7. The _____ of a satellite signal is expressed in decibels (dB) referenced to 1 _____ of power.

8. S-band satellites transmit using frequencies in the 2.6 _____ frequency band.

9. The _____ of a satellite antenna is a function of signal wavelength and antenna diameter.

10. The _____ of a communications signal can be determined by dividing the speed of light by the signal's frequency.

Part II: True or False

Mark each statement below "T" (true) or "F" (false) in the blank provided.

____ 11. The apparent spacing in degrees between two adjacent geostationary satellites is always smaller than the actual spacing between the two spacecraft in degrees of longitude.

____ 12. "Extended" C-band transponders operate within the 3.45–3.7 GHz and 4.2–4.8 GHz frequency ranges.

____ 13. As frequency increases, antenna beam width decreases.

____ 14. As frequency increases, wavelength decreases.

_____ 15. The beam width of a 1.2-m antenna is narrower when receiving S-band signals than it is when receiving Ku-band signals.

_____ 16. The apparent spacing between two geostationary satellites is always less than the spacing in degrees of longitude from one orbital location to the next.

_____ 17. The super high frequency bands are located between 2.5 GHz and 22 GHz.

_____ 18. Satellite transponder bandwidths typically range from 24 kHz to 108 kHz.

_____ 19. Most geostationary communication satellites transmit signals using two orthogonal senses of polarization so that each satellite can reuse the available satellite spectrum twice.

Part III: Multiple Choice

Circle the letter—a, b, c, d, or e—for the choice that best completes each of the sentences that appear below.

20. Signals propagated within the 3–30 MHz frequency range are also known as:

    a. microwaves

    b. medium waves

    c. short waves

    d. super waves

    e. none of the above

21. Communication satellites downlink their signals within which of the following frequency bands:

    a. C-band

    b. Ku-band

    c. M-band

    d. (a and b)

    e. (a, b, and c)

22. The narrow corridor within which a satellite antenna looks up at the sky is also known as the antenna _____.

    a. bandwidth

    b. isolation

    c. wavelength

    d. beam width

    e. apparent spacing

23. The beam coverage pattern that can cover 42.4 percent of the Earth's surface from a geostationary orbital location is called the _____.

    a. spot beam

    b. hemispheric beam

    c. global beam

    d. semiglobal beam

    e. zone beam

24. The signal level that would be generated if the satellite transponder is operated at full power is called:

    a. EIRP

    b. dBW

    c. saturation

    d. bandwidth

    e. none of the above

25. The satellite spectrum that typically is used for the high-power transmission of direct-to home TV signals is called the:

    a. C-band

    b. Ka-band

    c. S-band

    d. Ku-band

    e. X-band

INTERNET HYPERLINK REFERENCE

List of Geostationary Space Stations by Orbital Position and Publication covers all present and future communications satellites. http://www.itu.int/itudoc/itu-r/space/snl/sect1_b_21798.htm

CALCULATION OF WAVELENGTH AND ANTENNA BEAM WIDTH

Formula for calculation of signal wavelength:

$\lambda = c/F$

where λ = wavelength in meters, c = speed of light (300,000,000 meters/s), and F = frequency.

Thus, the wavelength of 4,200 MHz (4.2 GHz) is:

$\lambda = 300,000,000/4,200,000,000$
$= 0.0714$ m or
$\quad$ 7.14 cm

The beam width or beam angle θ (at the minus 3 dB points of the antenna radiation pattern) of a parabolic antenna is calculated in degrees using the above formula to calculate the wavelength in meters plus the following formula:

$\theta\ (-3\ dB) = 233\lambda/\pi D$

where: λ = wavelength in meters, π = 3.14159, and D = the diameter of the paraboloid in meters.

For example, the $\theta(-3\ dB)$ for a 2.4-m antenna at 4,200 megahertz would be:

$\theta(-3\ dB) = (233 \times 0.0714)/$
$\qquad (3.14159 \times 2.4)$
$\theta(-3\ dB) = (16.6362)/(7.5398)$
$\theta(-3\ dB) = 2.20645$ degrees

Chapter 2

Digital Video Compression Overview

The purpose of this chapter is to provide a general overview that contains more detailed information on those aspects of compression technology that professional installers and technicians are likely to encounter in the course of their normal work activities. Additional references are provided at the end of this chapter for those readers who would like to expand their knowledge on the subject to include other aspects that are of concern to broadcast engineers and system designers.

Digital video compression is the driving force behind the global revolution in satellite-delivered direct-to-home (DTH) TV program distribution. More than 23 million homes around the world were watching digital television by the end of 1998, with the vast majority of households receiving digital programs via satellite. These digital television signals are transmitted in an abbreviated format that dramatically reduces the amount of frequency bandwidth required without substantially degrading the quality of the received pictures and sound. The introduction of compression technology is causing a dramatic decline in the operational costs for TV service providers. The result has been a global explosion in the number of new satellite-delivered DTH TV services, including news, sports, movies, pay-per-view (PPV) special events, educational programming, and "narrowcast" broadcasts that can target the needs of small segments within any potential viewing audience.

Personal computers use digital compression techniques to reduce the amount of data storage capacity needed to save large computer files. For the past decade, the global telephone industry also has been using compression techniques to reduce the bandwidth—and consequently the cost—required for establishing narrow-band telephone circuits. During the early 1990s, communications engineers began developing very large-scale integrated circuits (VLSICs) and sophisticated software routines that could compress broadband telecommunications signals such as video.

FROM ANALOG TO DIGITAL TELEVISION

Radio and TV signals are made up of electromagnetic waves that continuously vary in their levels of frequency and/or intensity. All of these signals are called analog because of the wide signal variance that occurs within any transmission.

Digital transmission systems convert the visual and audio information into streams of binary digits or bits, strings of zeros and ones that correspond to the "off" and "on" logic states of computer circuitry. Any unique string of binary digits can be used to represent whatever we want it to be, as long as the receiving system understands the code words that we are using to convey the information. Digital standards in use around the world, such as ASCII (text) and GIFF (graphics), convert information to digital bit strings that are universally understood by all electronic systems receiving these digital signals.

To comprehend how digital video compression works, we must first have a working knowledge of the basic elements of analog TV technology. A conventional PAL or SECAM video signal contains 625 lines per individual image or frame. Flashed at a rate of 25 frames per second, each frame has two "interlaced" fields—each consisting of 312.5 lines—with field "1" displaying the odd-numbered lines and field "2" displaying the even-numbered lines (Figure 2–1). The interlaced scanning of the two fields occurs so rapidly that the eye perceives each complete image or "frame" rather than either separate field within any one frame.

Not all of the lines transmitted in each frame are actually displayed on the TV screen. For a PAL TV signal, only 576 lines out of a total of 625 are active lines; for NTSC, only 488 lines out of a total of 525 are active.

The cathode ray tube, which the TV set uses to display images, contains an electron gun that shoots a stream of electrons at the phosphor coating on the inside surface of the

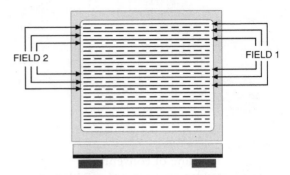

Figure 2–1 Each frame of PAL video has two interlaced fields, each consisting of 312.5 lines, which alternate at a rate of 50 times per second or 50 Hz.

TV screen. When the electron gun reaches the end of one line of video, a sync pulse is generated that switches off the electron gun so that it can move from screen right to screen left to begin "painting" the next active line of video onto the screen. The period at the end of each line during which the CRT's electron gun is deactivated is called the horizontal blanking interval (Figure 2–2).

At the end of either field, the CRT's electron gun will reach the end of the last line of active video. At this point, the electron gun must once again be disabled so that it can move from bottom screen right to top screen left in order to begin tracing the first line of the next field onto the TV screen. This inactive pe-

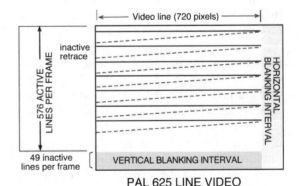

Figure 2–2 The blanking interval in each video frame turns off the CRT's electron gun so that it can move across the screen to begin the next line or field.

riod at the end of each field is called the vertical blanking interval.

Both the horizontal and vertical blanking intervals can be used to transmit data that is unrelated to the TV picture information. Many TV service providers, for example, use the vertical blanking interval to transmit teletext, test signals, conditional access data, and other information.

A single line of conventional analog PAL or SECAM video is composed of 720 picture elements or pixels. With 576 active lines in a single PAL video frame, the total number of pixels is therefore 720 × 576 or 414,720 pixels. Since PAL uses a screen refresh rate of 25 frames per second, 10,368,000 pixels are being sent to the TV screen each and every second.

BIT RATES

The amount of data information being transmitted in one second of time is called the bit rate, expressed in bits per second (b/s). A bit rate of one thousand bits per second is called a kilobit per second (kb/s); one million bits per second a megabit per second (Mb/s); and one billion bits per second a gigabit per second (Gb/s).

A bit rate of more than 200 Mb/s would be required to digitize a broadcast-quality video service without any signal impairment. This would require the use of several satellite transponders to relay just one uncompressed digital video signal. It therefore is essential that some form of signal compression be used to dramatically reduce the number of bits required for digital TV transmissions.

THE MOVING PICTURES EXPERTS GROUP

In 1988, the International Standards Organization (ISO) of the International Telecommunication Union established the Moving Pictures Experts Group (MPEG) to agree on an internationally recognized standard for the compressed representation of video, film, graphic, and text materials. The committee's goal was to develop a relatively simple, inexpensive, and flexible standard that put most of the complex functions at the transmitter rather than the receiver. Representatives from more than 50 corporations and governmental organizations worldwide took part in the MPEG committee's deliberations.

In 1991, the MPEG-1 standard was introduced to handle the compressed digital representation of nonvideo sources of multimedia at bit rates of 1.5 Mb/s or less. However, MPEG-1 can be adapted for the transmission of video signals as long as the video material is first converted from the original interlaced mode to a progressively scanned format, which is subsequently transmitted at half the normal field rate. MPEG-1 commonly is encountered on IBM computers and other compatible platforms with the ability to display files using the *.mpg extension. A few TV programmers initially elected to use a modified form of MPEG-1 called MPEG-1.5 to transmit via satellite while the MPEG committee developed a standard for source materials using interlaced scanning. Although not an official standard, MPEG-1.5 was adopted for use for a wide variety of applications, including the transmission of educational TV services and niche-program channels.

The MPEG committee selected its final criteria for a new standard in 1994 that resolves many of the problems with MPEG-1. The MPEG-2 standard features higher resolution, scalability, and the ability to process interlaced video source materials. MPEG-2 also features a transport stream that allows multiple video, audio, and data channels to be multiplexed at various bit rates into a single unified bitstream. There are so many similarities between MPEG-1 and MPEG-2, however, that MPEG-1 should be regarded as a subset of the MPEG-2 specification.

MPEG-2 COMPRESSION TECHNIQUES

MPEG compression is accomplished through the use of four basic techniques: preprocessing, temporal prediction, motion compensation, and quantization coding. Preprocessing filters out nonessential visual information from the video signal—information that is difficult to encode, but not an important component of human visual perception. Preprocessing typically uses a combination of spatial and temporal nonlinear filtering.

Motion compensation takes advantage of the fact that video sequences are most often highly correlated in time—each frame in any given sequence is very similar to the preceding and following frames. Compression focuses on coding the difference between frames rather than the encoding of each frame in isolation. Moreover, many of the changes that occur from frame to frame can be approximated as translations involving small regions of the video image. To accomplish this, an encoder scans subsections within each frame—called macroblocks—and identifies which ones will not change position from one frame to the next.

The encoder also identifies predictor macroblocks for those portions of a scene that are in motion, while noting their direction of motion and speed of movement. Only the relatively small difference, called the motion compensated residual, between each predictor block and the affected current block is subsequently transmitted to the integrated receiver/decoder or IRD. The IRD stores the information that does not change from frame to frame in its buffer memory. The IRD accesses the static macroblocks (Figure 2–3) from the buffer memory and then uses the motion compensation residual information to fill in the blanks.

The main drawback to the use of motion compensation is that motion artifacts will occur whenever an insufficient number of bits are available to describe detailed and/or

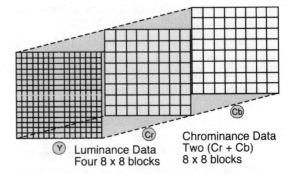

Luminance Data
Four 8 x 8 blocks

Chrominance Data
Two (Cr + Cb)
8 x 8 blocks

Figure 2–3 A macroblock in a 4:2:0 sampling format consisting of four 8 × 8 pixel blocks carrying the luminance or brightness information and two 8 × 8 blocks carrying the chrominance or color information.

rapidly changing scenes. These artifacts are most obvious when observing live video of sports events. The only way to overcome motion artifacts is for the programmer to increase the bit rate assigned for the transmission of the sports service.

A mathematical algorithm called the discrete cosine transform (DCT) reorganizes the residual difference between frames from a "spatial" domain into an equivalent series of coefficient numbers in a "frequency" domain that can be more quickly transmitted. The DCT is a trigonometric formula derived from Fourier analysis theory that can dramatically minimize the duplication of data within each picture. In the frequency domain, most of the important high-energy picture elements are represented by low frequencies at the top left corner of the block; less important visual information is represented as higher frequencies at the bottom right of the block.

Quantization coding converts the resulting sets of coefficient numbers into even more compact representative numbers by rounding off or scaling all coefficient values, within a certain range of limits, to the same value (Figure 2–4). For example, the quantization process is weighted so that the high-frequency areas in each block, for which the human eye is less sensitive, are reduced. Although this

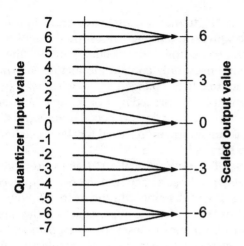

Figure 2–4 Quantization coding scales all coefficient values, within a certain range of limits, to the same value.

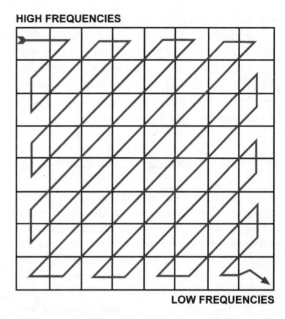

HIGH FREQUENCIES

LOW FREQUENCIES

Figure 2–5 The DC coefficients are scanned in a zigzag pattern that orders the results in descending values from top left to bottom right.

scaling process results in an approximation of the original signal, it is close enough to the original to be acceptable to the human eye.

The MPEG-2 digital bitstream is created by scanning the 64 frequency coefficients in a zigzag fashion from top left to bottom right (Figure 2–5). This results in the less important high-frequency areas being represented by strings of zeros. Spatial compression is achieved by coding the number of zeros in a row rather than coding each zero by itself. This is called "run-length coding."

The encoder also refers to an internal index, or codebook, of possible representative numbers from which it selects the code word that best matches each set of coefficients. This process is often compared to Morse code. With Morse code, the shortest code (.) is assigned to the letter "E," which is the most frequently occurring letter in the English alphabet; long codes are assigned to letters, such as Q (- - . -) and Z (- - . .), that occur in the English language less frequently. Quantization assigns short code words to those events that have a greater probability of occurring and long code words to coefficients that are less probable. This final stage of the quantization process is called "variable length coding."

GROUP OF PICTURES

The compression process divides each group of pictures making up any video scene into smaller segments, which are then subjected to coding (Figure 2–6). The group is first divided into individual video frames. Different frame

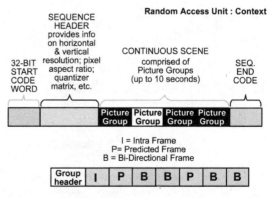

Figure 2–6 A group of pictures is a video sequence consisting of a series of interrelated frames.

options are available to the programmer. Full-resolution PAL video consists of an array of 720 pixels by 576 active lines. Lower resolution arrays, which consist of 720 pixels by 288 active lines or 360 pixels by 288 active lines, also may be used to reduce the number of bits required to transmit a video signal.

I, P, AND B FRAMES

MPEG-2 employs temporal prediction to minimize the duplication of data contained within any group of pictures by reducing selected elements within each frame into compact motion vector data as well as transmitting other data that describe the difference between one frame of video and any other. To accomplish all of this, three distinct frame types are used within any group of pictures. These are called the I, P, and B frames (Figure 2–7).

The intra-frame, or "I" frame, serves as a reference for predicting subsequent frames. I frames, which occur on an average of one out of every 10 to 15 frames, are essential for maintaining program continuity. The compression that occurs within an I frame only relates to the information contained within this frame. Each group of pictures must begin with an I frame. The regular insertion of I frames within the data stream is controlled by the encoder.

"P" frames are predicted from information presented in the nearest preceding I or P frame. Compression occurs because the P frame need only contain the picture information that has changed from what was presented in the preceding I or P frame. A buffer memory circuit in the decoder provides the missing information stored for the preceding I or P frame.

The bidirectional "B" frames are coded using prediction data from the nearest preceding "I" or "P" frame *and* the nearest following "I" or "P" frame. The encoder selects the number of B frames to insert between pairs of I or P reference frames, as well as selecting the most efficient overall sequence order. Although a more efficient level of compression is achieved by "B" frame usage, compatible receiver/decoders must have an additional memory buffer, which increases the cost of the decoder.

SLICES

Each video frame is also divided into segments called slices (Figure 2–8). With a PAL video signal, the 576 active lines are divided by 16 to create 36 slices. Each line is further subdivided (720/16 = 45) into 45 macroblocks (Figure 2–9).

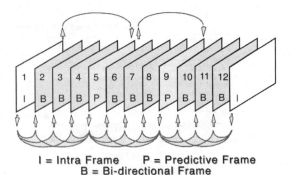

I = Intra Frame   P = Predictive Frame
B = Bi-directional Frame

Figure 2–7 Because of the high level of redundancy between frames in a group of pictures, only the picture information that changes from one frame to the next actually needs to be transmitted.

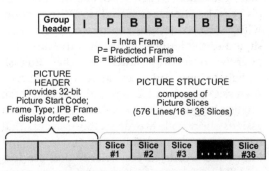

Figure 2–8 During MPEG video encoding, each frame of video is divided into units called slices.

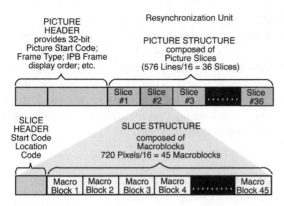

Figure 2–9 *Slices in turn are divided into smaller units called macroblocks that are subject to complex mathematical processes.*

MACROBLOCKS

Analog color TV cameras produce RGB (red, green and blue), or the luminance (Y) and color difference (R-Y and B-Y) signals. These component signals, which must be carried on three separate BNC cables, are susceptible to interference. The MPEG-2 macroblock consists of four blocks of brightness or luminance (Y) information, which together make up a 16 × 16 pixel array, as well as two or more 8 × 8 pixel blocks that represent the color or chrominance difference signals Cb and Cr. This macroblock configuration is referred to as a 4:2:0 format. With a 4:2:2 sampling format, four luminance and four chroma blocks, two representing Cb and the other two Cr, are employed; a 4:4:4 format contains four luminance and eight chroma blocks, four representing Cr and four representing Cb.

BLOCKS

A complex mathematical process called the discrete cosine transform (DCT) removes the spatial redundancy that occurs within each block. DCT coefficients are applied to each of these blocks to convert the luminance and chrominance picture information from a spa-

tial to a frequency domain (Figure 2–10). This transformation results in the creation of a matrix block consisting of DCT coefficients that effectively represent the intensity pattern within the block. These coefficients are then subjected to quantization, where they become one of a limited number of integer values that can be transmitted using fewer bits (Figure 2–11). Quantization takes advantage of the limitations of human perception by converting an infinite range of values to a limited set that corresponds to the visual response of the human eye. A nonlinear scaling process is used to determine how each coefficient is quantized. Following quantization, most of the DCT coefficients will be equal to zero. Zigzag

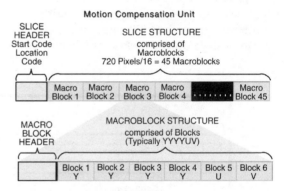

Figure 2–10 *Each block within a macroblock is submitted to the discrete cosine transform mathematical process for conversion from a spatial domain to a frequency domain.*

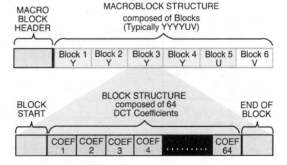

Figure 2–11 *The discrete cosine transform converts 8 × 8 blocks into a matrix of 64 DCT coefficients.*

scanning organizes the coefficients in order of frequency from top left to bottom right.

PACKETIZED ELEMENTARY AND TRANSPORT STREAMS

The MPEG-2 encoder compresses video and audio into packets. This results in a sequence of variable length packets being multiplexed into a packetized elementary stream (PES). The encoder uses a time field called the program clock reference (PCR) to synchronize the streaming of the video and audio packets. Programmers have the option of sending the PCR either inside a separate packetized elementary stream or as part of the same PES transporting the program information.

The MPEG-2 encoder then multiplexes numerous individual PES streams together to create a single unified MPEG-2 transport stream that shares common system information and teletext components. This unified transport stream, which consists of fixed-length packets 188 bytes in length, can contain numerous programs, each with an independent time base. Each transport packet is preceded by a transport header that includes information for bitstream identification.

MPEG-2 PROFILES, LEVELS, AND LAYERS

The MPEG-2 compression standard is actually a family of systems, with each system having an arranged degree of commonality and compatibility. MPEG-2 supports four different levels: High, High-1440, Main, and Low levels (Figure 2–12). The design for each level supports a variety of pixel arrays and frame rates. The High and High-1440 levels can support

MPEG-2 PROFILES: (Note: DVB does not support the SNR & Spatial Profiles)

MPEG-2 LEVELS:	Spatial Resolution Layer	Simple	Main	SNR	Spatial	High
HIGH 80 Mbit/s maximum	Enhancement		1920x1152x25 1920x1080x30			1920x1152x25 1920x1080x30
	Base Layer					960x576x25 960x480x30
HIGH-1440 60 Mbit/s maximum	Enhancement		1440x1152x25 1440x1080x30		1440x1152x25 1440x1080x30	1440x1152x25 1440x1080x30
	Base Layer				720 x 576 x 25 720 x 480 x 30	720 x 576 x 25 720 x 480 x 30
MAIN 15 Mbit/s maximum	Enhancement	720 x 576 x 25 720 x 480 x 30	720 x 576 x 25 720 x 480 x 30	720 x 576 x 25 720 x 480 x 30		720 x 576 x 25 720 x 480 x 30
	Base Layer					352 x 288 x 25 352 x 240 x 30
LOW 4 Mbit/s maximum	Enhancement		352 x 288 x 25 352 x 240 x 30	352 x 288 x 25 352 x 240 x 30		
	Base Layer					

Figure 2–12 MPEG-2 Profiles, Levels, and Layers.

high-definition TV (HDTV) and advanced definition TV (ADTV) pictures with 1,920 × 1,152 and 960 × 576 pixel arrays, respectively. Both the Main level and the Low level can support standard TV pixel arrays of 720 × 576 or 352 × 288. All but one level support two spatial resolution layers, called the Enhancement Layer and the Base Layer.

All digital bitstreams and set-top boxes also are classified according to video frame rate, either 25 Hz or 30 Hz, depending on the accepted standard in each country of operation. Set-top boxes with dual frame rate capabilities are also possible. Although digital bitstreams are set for one of the two frame rates, it also is possible for an MPEG-2 transport stream to relay program material for more than one type of IRD.

MPEG-2 also supports five different Profiles: Simple, Main, SNR Scalable, Spatial Scalable, and High (Table 2–1). Each Profile consists of a collection of compression tools. For example, the Main Profile may use up to 720 pixels per line at Main Level, or up to 1,920 pixels per line at High Level. Most 525- and 625-line broadcast TV signals use the Main Profile at the Main Level (MP@ML), whereas most future

HDTV signals or ADTV signals will use the High Profile at either the High Level or the High-1440 Level.

MPEG-2 achieves a high degree of flexibility by incorporating two spatial resolution layers for each of the available Levels and Profiles previously described. A single MPEG-2 transport stream can simultaneously deliver standard TV as well as ADTV or HDTV signals in an economical fashion. This is accomplished by using the low-resolution Base Layer to deliver a standard TV signal, at the same time using one or more Enhancement Layers to deliver the additional data required to produce higher resolution TV pictures. Together the enhancement and low-resolution layers deliver all the information that the HDTV set needs to produce a high-resolution picture. Standard TV sets receive the data they require exclusively from the Base Layer, while ignoring the data contained in the Enhancement Layer.

MPEG-2 transport streams that only use one layer are called nonscalable digital video bitstreams; those supporting two or more layers are called scalable hierarchies. Those transport streams with scalable hierarchies offer the added benefit of producing a more robust signal that is less prone to transmission path errors.

MPEG-2 ENCODING RATES

A unified MPEG-2 digital bitstream or multiplex may contain eight or more TV services with associated audio, auxiliary audio services, conditional access data, and auxiliary data services, such as teletext or Internet connectivity. A single video signal within this bitstream will invariably have a lower bit rate. For example, a single VHS-quality movie can be transmitted at a bit rate of 1.5 Mb/s; a news or general entertainment TV program at 3.4–4 Mb/s; live sports at 4.6–6 Mb/s; or studio-quality broadcasts at a rate of more than 8 Mb/s (see Figure 2–13). The encoding rate required for any MPEG-2 broadcast varies according to

Table 2–1 MPEG-2 Profiles Chart

Simple Profile	The MPEG-2 profile with the fewest available tools.
Main Profile	Contains all of the tools offered by the Simple Profile plus the ability to interpret B frames for bidirectional prediction purposes.
SNR Scalable and Spatial Scalable Profiles	Adds tools that allow the video data to be partitioned into a base layer and one or more enhancement layers, which can be used to improve video resolution or the video signal-to-noise ratio (SNR). The DVB standard does not support any of the SNR or Spatial Scalable Profiles offered in the MPEG-2 specification.
High Profile	Contains all of the tools offered by the other Profiles plus the ability to code line-simultaneous color-difference signals.

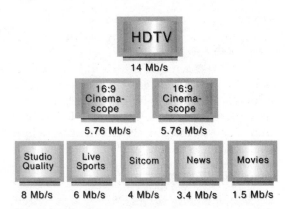

Figure 2–13 Typical MPEG-2 transmission rates in megabits per second.

the bit allocation decisions made by each program service provider.

The MPEG-2 encoder located at the satellite uplink has a finite amount of time to make encoding decisions (Figure 2–14). Live sports and other live action materials require a higher data rate because the encoder is forced to make immediate coding decisions and must also transmit complex, rapid motion changes without introducing high levels of distortion.

SATELLITE TRANSMISSION FORMATS

Most digital DTH satellite TV programmers use a transmission format called multiple channel

per carrier (MCPC) to multiplex two or more program services. With MCPC, a package of program services can use the same conditional access and forward error correction systems, thereby economizing on the overall bandwidth and transmission speed requirements (Figure 2–15).

Moreover, programmers can dynamically assign capacity within the digital bitstream of any multiplexed transmission, so that more bits are available to a live sport broadcast and fewer bits to a news report or interview program consisting of "talking heads." At the conclusion of a live basketball game, for example, the digital capacity previously used to relay a single sport event even could be used to simultaneously transmit two or more separate movie services.

MCPC systems employ a transmission technique known as time division multiplex (TDM). With TDM, multiple programs are assigned different time slots within a designated time frame and transmitted in bursts at a high bit rate. The digital IRD selects the packets of information for the service that it is tuned to receive while ignoring and discarding all other packets being transmitted over the transponder. In this way, each program within the multiplex has access to the entire transponder power and bandwidth.

Video Services	Data Rate
High Definition Television (HDTV)	14.0 Mb/s
Studio Quality CCIR 601	8.064 Mb/s
16:9 Wide Screen Aspect Ratio	5.760 Mb/s
Live Sports	4.608 Mb/s
Film/Broadcast	3.456 Mb/s
Pay Per View Movies	1.152 Mb/s
Musicam Audio	
Monaural	128 kb/s
Stereo	256 kb/s
Stereo Pair	512 kb/s
Digital Data	9.6 kb/s
Service Control Data	30.72 kb/s
Component ID Overhead (total data rate)	2 percent

Figure 2–14 Recommended MPEG-2 data rates (minimum).

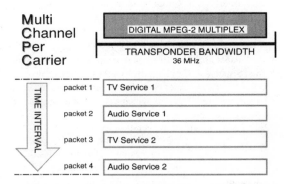

Figure 2–15 MCPC multiplexes all video, audio, and data into a single unified bitstream. All data are delivered sequentially over time in uniform-sized packets.

Until recently, the major drawback to MCPC was that all video, audio, and data source materials had to be present at the master satellite uplink facility that generated the MPEG transport stream. Newly launched satellites such as Hot Bird 4, however, carry onboard multiplexers that can generate an MPEG-2 transport stream. This allows program source materials to be uplinked to the satellite from multiple locations.

A few satellite TV services use an alternate digital transmission format called single channel per carrier (SCPC), which allows multiple uplink facilities to transmit feeds to a satellite transponder from separate locations (Figure 2–16). SCPC most often is used for specialized applications—such as satellite newsgathering (SNG) or educational TV—where it is difficult or even impossible for all program source materials to be uplinked from a single location. SCPC-based digital services typically use a fraction of a transponder rather than its entire bandwidth. When multiple SCPC services share the same transponder, each service is transmitted individually on its own narrow-band frequency carrier within the transponder (Figure 2–17). Guard bands and power back-off also must be employed between the SCPC carriers to prevent interference between services.

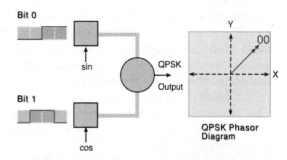

Figure 2–17 DVB-compliant satellite transmissions use quadrature phase shift keying modulation. With QPSK, each signaled state encodes a two-bit symbol.

MPEG-2 SATELLITE MODULATION TECHNIQUES

MPEG-2 satellite transmissions use a digital modulation technique known as QPSK (for quadrature phase shift keying). The production of QPSK modulation requires the simultaneous processing of two bits of information, whereby the data rate is effectively doubled without a corresponding increase in the signaling rate (Figures 2–17 and 2–18).

A simple modulation system such as binary phase shift keying (BPSK) varies the carrier frequency between two distinct phase states to correspond to the binary digits 1 (on) and 0 (off). QPSK, however, uses four phase

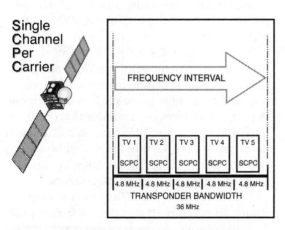

Figure 2–16 SCPC places the video, audio, and data for each TV service within discrete segments of the satellite transponder.

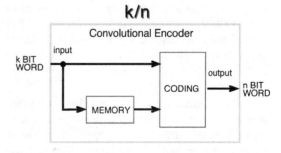

Figure 2–18 DVB-compliant digital bitstreams contain special codes that the IRD can use to check to ensure that all bits of information sent have actually been received.

states instead of two, allowing two bits to be transmitted simultaneously by switching between different combinations—called symbols—of the four available states. The digital encoder at the uplink converts bits in pairs (called "di-bits") into equivalent two-bit symbols. The symbol rate for compressed digital video transmissions is expressed in megasymbols per second (Msym/s).

The symbol rate for any given digital TV program stream must be programmed into the IRD before reception can occur. This symbol rate usually varies from one QPSK-modulated stream to the next. Digital IRDs most always are preprogrammed at the factory to automatically tune to the symbol rate used by a particular digital TV service provider. Those TV viewers who intend to receive multiple bouquets using the satellite receiving system will need to change the symbol rate for the IRD whenever switching from one program bouquet to another.

THE DIGITAL VIDEO BROADCASTING STANDARD

Most digital satellite TV broadcasters today are using a version of MPEG-2 that conforms to parameters adopted by Europe's Digital Video Broadcasting (DVB) Group. The formal DVB standard, which was first published in January 1995 by the European Telecommunication Standards Institute (ETSI), has since been adopted by other broadcast entities around the world (see Table 2–2). Moreover, the DVB Group has selected the MPEG-2 Main Profile at Main Level (MP@ML) with a maximum data rate of 15 Mb/s as the basis of its digital compression system.

What makes the DVB standard a significant development is its ability to serve as a unified standard that can be applied across a variety of distribution platforms. Many of the same elements are used in the various DVB systems presented below to permit the distribution of signals among different platforms

without the need for complex and costly MPEG-2 decoding and recoding.

DVB allows MPEG-2 digital signals to be seamlessly transported between various satellite (DVB-S), cable (DVB-C) terrestrial TV (DVB-T), SMATV (DVB-CS) and MMDS (DBV-MC or DVB-MS) distribution platforms without requiring any modification to the original transport stream. Satellite-delivered signals, for example, can be demodulated at cable head ends or terrestrial broadcast facilities and then seamlessly remodulated for distribution. This dramatically streamlines the transferral process and results in economies of scale that would not have been realized if mutually incompatible systems had won the day.

The MPEG-2 encoder (Figures 2–19 and 2–20) multiplexes all data into packets, with each packet containing a 1-byte header and a 187-byte message. The header—which contains the packet identifier (PID)—provides the instructions that the IRD needs to know what to do with the message contained in each packet. For example, the IRD only needs to process those packets that contain the information pertaining to the service that the IRD is set to receive. All other packets in the digital transport stream can be ignored and discarded. Four different packet identifiers commonly are available. The VPID is the packet identifier for video data, and the APID is the packet identifier for audio data. The digital transport stream also must send a program clock reference (PCR PID) at intervals that the IRD uses to synchronize the VPID and APID packets. A data packet identifier (DPID) also is required to identify those packets that contain auxiliary data services, conditional access (CA) data, and the Service Information and Teletext data, which may include the digital bouquet's satellite transmission frequencies, channel allocations, and modulation parameters.

The value of the DVB standard's Service Information (DVB-SI) component is that it renders any changes to the bouquet configuration transparent to the end user by permitting each programmer to reconfigure the IRD software

Table 2–2 DVB Specifications Chart

DVB-S	A digital satellite broadcasting system for television, sound, and data services that predominantly downlinks in the 11/12 GHz frequency spectrum. DVB-S includes specifications governing framing structure, channel coding, and QPSK modulation at 2 bits per symbol. DVB-S features modem standards for variable transponder bandwidths and data rates so that each broadcaster can match the transmissions to the available transponder bandwidth. DVB-S also supports two forward error correction methods: an outer FEC using Reed–Solomon block coding [204, 188, T=8] and an inner FEC that uses convolutional coding, with 35% half-Nyquist filtering and rates of 1/2, 2/3, 3/4, 4/5, 5/6 or 7/8.
DVB-C	The DVB-C cable specification is based on DVB-S, but the modulation scheme is Quadrature Amplitude Modulation (QAM) rather than QPSK. DVB-C is a cable broadcasting standard for television, sound and data services using standard cable TV distribution frequencies. The system is centered on QAM modulation with 64 symbols (64-QAM). However, lower-level systems, such as 16-QAM and 32-QAM, as well as higher-level systems such as 128-QAM and 256-QAM, also are available for use. DVB-C over an 8-MHz channel can accommodate a payload capacity of 38.5 Mbit/s if 64-QAM is used as the modulation scheme. The level of noise immunity varies as a trade-off of system capacity against the robustness of the data.
DVB-T	Approved in February 1997, DVB-T is a digital terrestrial broadcasting standard for television over standard broadcast TV frequencies. DVB-T uses a transmission scheme based on Coded Orthogonal Frequency Division Multiplexing that uses a large number of carriers to spread the information content of the signal. The main advantage of DVB-T is that it offers a very robust signal in a strong multipath environment. Because of the advantage of multipath immunity, a network of transmitting stations with overlapping coverage areas can operate on a single frequency.
	DVB-T is designed to use either 1,705 (2k) or 6,817 carriers (8k). Each carrier system uses QAM modulation with 4–64 symbols and 8 MHz of bandwidth. The 2k mode is applicable for single transmitter systems or relatively small single-frequency networks with limited transponder power. The 8k mode also can be used for single transmitter systems, but is more appropriate for large-area single frequency networks. Like DVB-S, DVB-T uses Reed–Solomon outer coding and convolutional inner coding for forward error correction purposes.
DVB-CS	This is a specification for satellite master antenna television (SMATV) systems that distribute programming to households located in one or more adjacent buildings. A common satellite dish is used to receive the signals, which are combined with terrestrial TV channels and then sent to each household by means of a cable distribution system. In this case, the SMATV head end is totally transparent to the incoming digital multiplex, which is delivered to each IRD in the system without any baseband interfacing required.
DVB-MC	This specification is for use with a multipoint distribution system (MDS) using digital technology and microwave frequencies. DVB-MC is based on the DVB-C specification for cable TV systems and therefore can use the same set-top box that digital cable systems use.
DVB-MS	This is another specification for use with multipoint distribution systems. The DVB-MS specification, however, is based on the DVB-S specification for satellite TV systems and therefore can use the same digital IRD that digital DTH systems use. Instead of using a satellite dish, the IRD is equipped with a small MDS antenna and frequency converter.
DVB-SI	The specification for Service Information (SI) in digital video broadcasting systems.
DVB-CI	The Common Interface Specification for conditional access and other digital video broadcasting encryption applications.
DVB-TXT	A specification for conveying teletext in digital video broadcasting applications.

automatically from the uplink. The digital IRD need only be set up once, usually preprogrammed at the factory, to find the first satellite transponder. After that, the IRD will be able to download all of the required transmission parameters, even if the programmer changes them from the original factory settings at a later date. The DVB-SI component also sets the parameters for the transmission of an electronic program guide (EPG) that can provide a wide variety of information, including service provider and channel name;

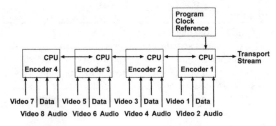

Figure 2–19 MPEG-2 encoder block diagram.

Table 2–3 Program Association Table

PAT (PID 0000) = 0100, 0200, 0300, 0400

PMT 1 (PID 0100) = Video PID 0101, Audio PID 0102,
Audio PID 0103, PCR 01FF

PMT 2 (PID 0200) = Video PID 0201, Audio PID 0202,
Audio PID 0203, PCR 02FF

PMT 3 (PID 0300) = Video PID 0301, Audio PID 0302,
Audio PID 0303, PCR 03FF

PMT 4 (PID 0400) = Video PID 0401, Audio PID 0402,
Audio PID 0403, PCR 04FF

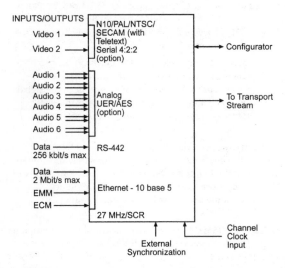

Figure 2–20 MPEG-2 encoder detailed view.

program name, type, and description; alternate channel program lists; and forthcoming program information.

The Service Information component contains a Program Association Table or PAT that provides each IRD with a list called a Program Map Table (PMT) that identifies every signal component within the MPEG-2 transport stream. An example of a PAT might look like the one in Table 2–3.

From this PAT, the IRD can determine that the transport stream contains four video services and corresponding stereo audio pairs, as well as separate timing information for each service. The PAT also will provide the IRD with other information, such as the name and duration of each program service, as well as any auxiliary data services that may be part of the digital bitstream.

A Network Information Table (NIT) also is available that provides the IRD with a list of the bouquet's associated transponders along with the transmission parameters for each transponder. In most instances, the associated transponders will be on the same satellite as the transponder to which the IRD is tuned. Some digital DTH systems, however, are equipped with a motorized antenna actuator that allows the IRD to receive signals from multiple satellites. In this case, the Network Information Table can supply the information that the IRD needs to locate associated transponders on other satellites.

The DVB-SI component also contains a Bouquet Association Table (BAT) that provides each IRD with comprehensive information about the program resources that are contained within the MPEG-2 transport stream. For example, the BAT can identify program content by category or theme. A separate Event Information Table (EIT) contains scheduling information as to when each program will air and for how long; the Time and Date Table (TDT) provides the IRD with the correct time and date.

DVB COMPATIBILITY ISSUES

Various digital satellite TV broadcasters and manufacturers emblazon logos onto their products that proudly proclaim that their units are "DVB-compliant." Does this mean that we have finally entered a new era of global video compatibility? We wish.

Although the DVB Group adopted a common interface for conditional access (DVB-CI), the committee did not agree on any single encryption or CA standard. Each digital IRD must have a compatible conditional access module and smart card before it can successfully receive any encrypted digital bouquet.

In some instances, national authorities have taken steps to ensure that all digital DTH programmers operating within their borders use the same conditional access system (see Figure 2–21). In Spain, for example, the three leading bouquet operators all use the same conditional access system, which allows viewers to subscribe to any of the available services and use the same IRD to gain access, albeit with the assistance of as many as three smart cards. More often, however, there will be two or more mutually incompatible bouquets available within the same country or region, with each bouquet requiring its own proprietary IRD and smart card.

Does a DVB-compliant IRD offer superior performance over other types of digital set-top boxes? Don't bet the farm on it. The signal quality produced by any digital delivery system, whether DVB or something else, is largely a function of how many bits are assigned to any given transmission within the digital bitstream; what the subscriber receives can range from quasi VHS all the way up to HDTV. The final result is up to each broadcaster.

FORWARD ERROR CORRECTION

The QPSK-modulated satellite signal contains special codes that the IRD uses to check that all bits of information sent have actually been received. This forward (sent ahead to the receiver along with the original message) error correction (FEC) technique creates a very robust signal with substantial advantages over an uncoded digital transmission containing the same information content.

Early coding experiments compared the performance of coded versus uncoded digital messages. The results showed that a signal improvement of 3.3 dB over the performance of the original message could be obtained through the use of error correction techniques. In other words, a satellite link that would require an antenna 1.8 m in diameter to receive an uncoded digital message could use a much smaller dish to obtain the same level of performance from a coded version of the same digital message. What's more, experimenters found that an encoder that used two coding techniques in cascade (i.e., one feeding into the other) could generate additional performance gains.

The FEC "overhead" consists of redundant symbols that are added to the original message. Although this increases the overall transmission rate and bandwidth requirements, the redundant symbols accentuate the uniqueness of the message in a way that prevents channel noise from corrupting enough symbols to destroy its uniqueness. The decoder uses the FEC symbols to restore data reliability after the message has been received.

One type of FEC encoding, called the Verbiti code, is expressed as a ratio, such as 1/2, 3/4, or 7/8. The numerator indicates the number of original symbols entering the encoder and the denominator indicates the number of error-corrected symbols leaving the encoder. Therefore, an FEC of 7/8 means that for every seven symbols entering the encoder, eight symbols leave; in other words, there will be one error-correcting symbol out of every 8 symbols.

The other type of FEC encoding, called the Reed–Solomon code, adds redundant symbols to individual strings or blocks of binary digits. The encoder accomplishes this task by only looking at the symbols that make up each discrete string or block of digital bits. Reed–Solomon uses 188 bytes out of every block of 204 bytes for transmitting the original

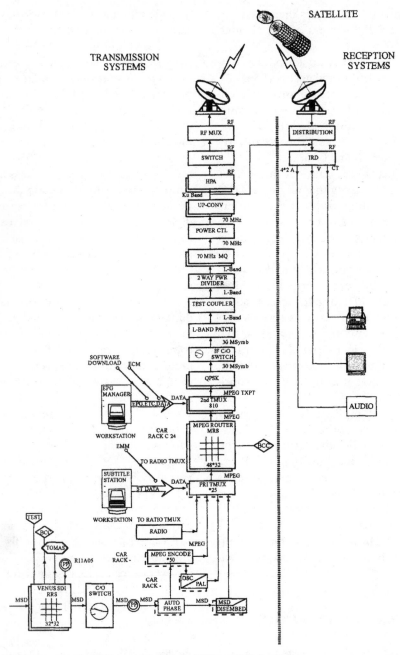

Figure 2–21 *Block diagram of a DVB-compliant digital DTH uplink facility.*

signal information. The remainder is used to send parity bits to the IRD to assist in the correction of any transmission errors.

The Reed–Solomon decoder uses an algorithm to simultaneously solve a set of algebraic equations based on the syndrome of parity checks from the retrieved block. It is particularly good at detecting and correcting bit errors generated by burst noise that can be caused by automobile ignition noise or microwave ovens operating in the general vicinity of the receiver.

FEC systems such as Verbiti, which look at previously transmitted blocks as well as the current block, are called convolutional coding systems. The convolutional encoder has a buffer circuit that holds previously coded messages in memory for reference. Convolutional coding is particularly effective in correcting or concealing thermal noise bit errors.

When speaking of FEC coding systems, concatenation is said to take place when the output of one encoder falls or cascades into a second encoder. The first code is referred to as the inner code; the second is called the outer code. MPEG-2 DVB-compliant systems use convolutional coding as the inner code, with coding rates of 1/2, 2/3, 3/4, 5/6, or 7/8 available for use, and Reed–Solomon block coding as the outer code.

DIGITAL BOUQUET TRADE-OFFS

As previously mentioned, the symbol rates and FEC rates in use often vary from one digital bouquet to the next. The obvious question is why. The maximum possible symbol rate is a function of satellite transponder bandwidth. This can be calculated by the following formula:

Maximum Symbol rate = BW/1.2

For example, a 33-MHz wide transponder/1.2 = 27.5 Msym/s.

Suppose a programmer uses an FEC rate of 3/4. The digital bit rate for the multiplex would be:

27.5 Msym/s × 2 (2 bits per symbol for QPSK modulation) = 55 Mbit/s

55 Mbit/s × 3/4 (the inner code FEC rate) = 41.25 Mbit/s

41.25 Mbit/s × 188/204 (the outer code Reed–Solomon FEC rate) = 38.015 Mbit/s

In another example we shall alter only the FEC rate from 3/4 to 1/2 to see how the change in the inner code FEC rate affects the available bits for signal transmission.

27.5 Msym/s × 2 (2 bits per symbol for QPSK modulation) = 55 Mbit/s

55 Mbit/s × 1/2 (the inner code FEC rate) = 27.5 Mbit/s

27.5 Mbit/s × 188/204 (the outer code Reed–Solomon FEC rate) = 25.343 Mbit/s

Although the 1/2 FEC rate will result in a very robust signal, it also will dramatically reduce the number of program services that the programmer can send over any given transponder. Each programmer must therefore make a decision on transmission parameters that balances the desirability of having a robust signal against the need to transmit as many program services as possible through each satellite transponder.

BIT ERROR RATE AND E_B/N_O

Measured in exponential notation, the bit error rate (BER) quantifies the performance level of the digital link. A BER of 1×10^{-3} expresses the probability of one bit error occurring in a block of 1,000 bits. A BER of 5.0×10^{-5} is superior to a lower BER of 9.0×10^{-4} because there is a probability that fewer bit errors will occur. BER also may be expressed as 5E-4 or 3E-3, which is equivalent to a BER of 5×10^{-4} or 3×10^{-3}.

The quantifying measure of a digital satellite link is E_b/N_o: the ratio of bit energy to noise density (Figure 2–22). QPSK modulation has the ability to achieve a given BER at a relatively low E_b/N_o when used for wide-band applications such as satellite communications.

The received E_b/N_o, which is expressed in decibels (dB), represents the signal-to-noise ratio of the receiving system. Another way to gauge the importance of E_b/N_o is to realize that

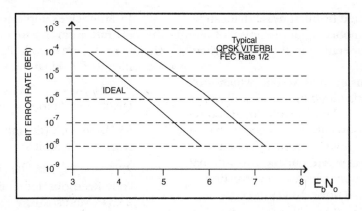

Figure 2–22 *The relationship of E_b/N_o to bit error rate (BER).*

as E_b/N_o increases, the number of bit errors decreases. Error correction is used to achieve a given BER at as small a value of E_b/N_o as possible. The DVB specification calls for a worst-case bit error probability of 1E-11. This equates to not more than one bit error in a 38 Mbit/s data stream every 45 minutes, or no more than one bit error in a 8 Mbit/s digital TV program service every 3.5 hours.

KEY TECHNICAL TERMS

The following key technical terms were presented in this chapter. If you do not know the meaning of any term presented below, refer back to the place in this chapter where it was presented or refer to the Glossary before taking the quick check exercises that appear below.

Analog

Bit

Bit Error Rate (BER)

Block

Chrominance

Digital

Discrete cosine transform (DCT)

DVB

E_b/N_o

Encoder, encoding rate

Forward error correction (FEC)

Horizontal blanking interval

Interlace scanning

Luminance

Macroblock

Motion compensation

MPEG, Profiles, Levels, and Layers

Multiple channel per carrier

Packetized elementary stream

Packets

Quadrature phase shift keying

Quantization

Single channel per carrier

Symbol rate

Time division multiplex (TDM)

Transport stream

Variable length coding

Vertical blanking interval

Video field and frame

QUICK CHECK EXERCISES

Check your comprehension of the contents of this chapter by answering the following questions and comparing your answers to the self-study examination key that appears in the Appendix.

Part I: Matching Questions

Put the appropriate letter designation—a, b, c, d, etc.—for each term in the blank before the matching description.

a. megabits per second (Mbit/s)

b. frame

c. field

d. MPEG-1

e. MPEG-2

f. interlace scanning

g. progressive scanning

h. multiplex

i. bit rate

j. luminance

k. chrominance

l. gigabits per second (Gb/s)

m. B frames

n. forward error correction (FEC)

o. motion compensation

p. bit error rate (BER)

q. motion-compensated residual

1. Digital video compression systems using _____ achieve a higher level of bit rate efficiency but require that the decoder possess a second buffer memory circuit that adds to the cost of the receiving system.

2. A single _____ of PAL video contains 625 lines.

3. Digital DTH satellites typically transmit digitally at bit rates of 1.5 to 8 _____ .

4. Digital satellite transmission systems usually use some form of _____ in order to improve the accuracy of the received data.

5. The "threshold" level of a digital satellite TV receiver is usually specified as a _____ .

6. The _____ digital video compression standard is used to process all _____ sources of media, such as film-based materials, text, and computer graphics.

7. Two essential components of any video signal are the black-and-white or _____ information as well as the color or _____ information.

8. _____ is used to compute the direction and speed of moving objects in a digitally compressed video image.

9. Digital satellite transmissions using MCPC combine numerous video, audio, and data signals into a single digital bitstream or _____ .

10. The small difference between each predictor macroblock and the affected current macroblock is called the _____ .

Part II: True or False

Mark each statement below "T" (true) or "F" (false) in the blank provided.

____ 11. The MPEG-2 standard was developed to handle the digital compression of all media using interlaced scanning techniques, such as broadcast TV signals.

____ 12. Convolutional and block encoding techniques are used by several Asian broadcasters to prevent unauthorized access to satellite TV signals that are transmitted in an analog format.

____ 13. MPEG compression relies on preprocessing techniques to remove those components of a video image that are not essential to human perception.

____ 14. The ultimate goal of every digital DTH installer is to fine-tune the antenna and feed in order to achieve as low a receiver bit error rate (BER) as possible. For example, a BER of 3×10^{-3} (3 E-3) would be vastly superior to a BER of 6×10^{-7} (6 E-7).

____ 15. Video resolution quality is determined by the number of active video lines being transmitted as well as the number of picture elements or "pixels" contain within each line.

____ 16. All satellite TV service providers that use digital video compression must multiplex all of their video, audio, and data signals into a single digital bitstream before uplinking to any given satellite.

____ 17. Variable length coding is often compared to the Morse code system because frequently transmitted messages are assigned the longest code sequences, while infrequently transmitted messages are assigned very short code sequences.

INTERNET HYPERLINK REFERENCES

North American MPEG-2 Information by Rod Hewitt.
http://www.coolstf.com/mpeg/

The New European Digital Video Broadcast (DVB) Standard by Markus Kuhn is a brief introduction into the technology behind the DVB/MPEG-2 digital TV broadcasting system.
ftp://ftp.informatik.uni-erlangen.de/pub/multimedia/tv-crypt/dvb.txt

What Is MPEG Digital TV? by Neil Anthony Powell at UK Satellite Control.
http://www.sat-net.com/uk-satellite/MPEG-Digital.html

A Concise Summary of the Digital Video Broadcasting Group's DVB Standards for Satellite Terrestrial and Cable TV Use.
http://www.dvb.org/dvb_framer.htm

The Leitch Web Site's Digital Television Tutorials. Video Compression (Strachan, February 1996), An Introduction to Digital Television (David Strachan, March 1995), Digital Signal Conversion (Strachan and Conrod, June 1995), and Error Detection and Handling in Digital Television (Strachan and Conrod, October 1995).
http://www.leitch.com/Reference/Reference.htm#Tutorials

Chapter 3

Digital DTH Platforms and Bouquets

More than 2,000 digital TV and radio channels currently are available from various geostationary satellite platforms worldwide. Moreover, more than 30 million households worldwide were expected to be receiving digital TV signals via satellite by the end of 1999. The extent to which any individual can directly access these channels will depend on the receiving site location, subscription availability, and the type of digital DTH receiving system in use. The purpose of this chapter is to profile the technical characteristics of the major digital satellite platforms and present the basic operating parameters for the major digital TV program packages, called bouquets, in operation at the time of writing.

ITU SATELLITE SERVICES

The International Telecommunication Union (ITU), the United Nations organization that is responsible for the registration and coordination of geostationary satellite communications systems on a worldwide basis, has established three general satellite service categories: fixed, mobile, and broadcast. Within the ITU's fixed satellite service (FSS) category, frequency segments and orbital locations are assigned on a regional basis for the transmission and reception of private point-to-point communications services. FSS satellite allocations, however, also may be used to transmit signals to the general public, as long as the national telecommunications authority that registers the FSS satellite system with the ITU has agreed to do so.

The ITU's mobile satellite service (MSS) assigns frequency segments and orbital locations on a regional basis for the relay of telecommunications signals to and from land mobile vehicles, ships at sea, airplanes, and fixed satellite earth stations. The ITU's broadcast satellite service (BSS) assigns frequency segments and orbital locations on a regional basis for the distribution of video, audio, and data signals directly to individual homes.

THE ITU'S BROADCAST SATELLITE SERVICE

In 1977, the ITU convened a World Administrative Radio Conference (WARC-77) for the purpose of establishing a broadcast satellite service standard for ITU Region 1 (Europe, Africa and the Middle East) and ITU Region 3 (Asia, Australia and the Pacific Rim). Under the WARC-77 plan, each country in Europe was allocated up to five transponders in either the upper (12.1–12.5 GHz) or lower (11.7–12.1 GHz) halves of the broadcast satellite service band. The WARC-77 plan also designated four orbital locations for European BSS operations (37, 31, and 19 degrees west longitude and 5 degrees east longitude). It also mandated 12 degrees of orbital separation between the na-

tional BSS systems to minimize potential interference problems between adjacent satellites. Moreover, orthogonal senses of circular (right- and left-hand) polarization were adopted to permit reuse of the limited spectrum available and to provide a measure of isolation between adjacent BSS satellite systems using identical transponder frequencies.

A 1983 Regional Administrative Radio Conference determined the appropriate BSS standards for ITU Region 2 (The Americas). The RARC-83 broadcasting plan for the Americas (see Figures 3–1 and 3–2) assigned specific transponder frequencies (12.2–12.7 GHz), polarization formats (right-hand or left-hand circular), and orbital locations to each country within the region. RARC-83 also assigned a minimum of 9 degrees of orbital separation between satellite systems serving adjacent or overlapping geographical areas.

Both WARC-77 and RARC-83 anticipated that only the most powerful satellites using

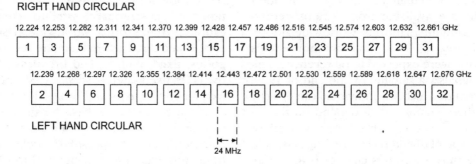

Figure 3–1 RARC-83 high power DBS (BSS) frequency assignments.

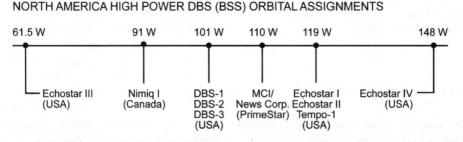

Figure 3–2 RARC-83 high power DBS (BSS) orbital assignments.

super high frequencies could transmit TV programming directly to individual homes. The technical advances that immediately followed, however, soon rendered the WARC-77 BSS plan obsolete. By the mid-1980s, satellite TV programmers in the United States and Europe began broadcasting their signals using satellite frequencies and orbital positions that the ITU had originally assigned for nonbroadcast, fixed satellite service (FSS) operations.

The technical characteristics of any satellite platform largely determine the performance capabilities of each digital DTH bouquet using that platform. These technical characteristics include the satellite's transponder output power, footprint, polarization format, and orbital spacing environment. Readers may use the EIRP-to-dish-size conversion charts presented in Figures 3–3 and 3–4 and the satellite footprint maps that appear throughout this chapter to quickly determine the appropriate antenna size for receiving a particular digital DTH bouquet from any given location.

The digital DTH service descriptions presented next provide the main parameters to which the receiving system's digital IRD must tune to receive the available program bouquets. These include the bouquet's digital transmission rate expressed in millions of symbols, or

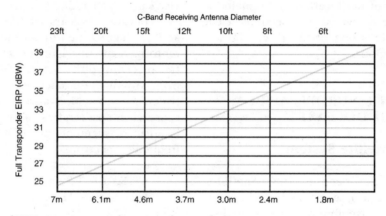

Figure 3–3 C-band EIRP versus antenna diameter reference chart.

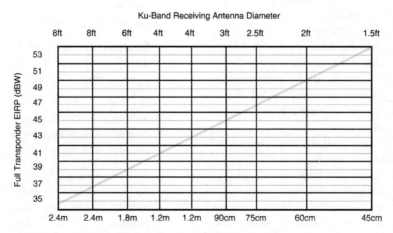

Figure 3–4 Ku-band EIRP versus antenna diameter reference chart.

megasymbols, per second (Msym/s), the forward error correction (FEC) rate, and, in the case of subscription TV services, the conditional access system in use. The Internet addresses for the major satellite operators and digital DTH service providers also are provided so that readers can use the World Wide Web (WWW) to obtain the latest information about new satellite platforms, transmission parameters, and digital program availability.

In today's complex satellite world, digital DTH programming is delivered by a multiplicity of satellite platforms, most of which do not conform to the original WARC-77 or RARC-83 broadcast satellite service plans. To assist the reader in locating the information that is most relevant to his or her location, the remainder of this chapter is divided into sections. Each section represents one of the three ITU administrative regions.

ITU REGION 1: EUROPE, AFRICA, AND THE MIDDLE EAST

The Astra Satellite System

The Socete Europeenne des Satellites (http://www.astra.lu), a privately held company based in Luxembourg, was created in 1985 to develop and operate the Astra satellite system. On December 10, 1988, SES's Astra 1-A satellite was launched by an Ariane 44P launch vehicle to an orbital assignment of 19.2 degrees east longitude. Equipped with 45-watt transponders, the new satellite soon began transmitting 16 channels of TV programming to cable TV systems and individual homes equipped with receiving antennas as small as 2 feet (60 cm) in diameter.

Since the launch of the first Astra satellite, seven additional satellites (Astra 1-B, 1-C, 1-D, 1-E, 1-F, 1-G, and 2A) have been deployed, five of which are colocated with Astra 1-A at 19.2 degrees east longitude. At the time of writing, four of these spacecraft were dedicated to the transmission of digital DTH ser-

vices: Astra 1-E (see Figures 3–5 and 3–6), 1-F (see Figures 3–7 and 3–8), and 1-G (see Figures 3–9 and 3–10), and Astra 2-A at the new SES orbital location of 28.2 degrees east longitude. All of these satellites deliver a nominal effective isotropic radiated power (EIRP) of 52 dBW within the center contour of their respective footprints. This equates to a receiving antenna aperture of 60 cm in diameter.

Each Astra satellite also transmits signals through four downlink beams that slightly differ from each other in terms of their respective coverage areas. Digital DTH programmers are assigned to the Astra satellite and beam that provides the best coverage of the portion of Western Europe in which the particular service's potential audience is located.

Because six Astra satellites are colocated at 19.2 degrees east longitude, no expensive motorized antenna actuator is needed to move the receiving dish to view TV programming from any of these satellites. Interference between Astra satellites also is nonexistent because each spacecraft is designed to transmit in adjacent segments of the Ku-band satellite frequency spectrum.

ASTRA 1-E, 1-F, and 1-G (11.7–12.75 GHz, Polarization: Linear).

Dutch-language digital bouquet: ASTRA 1-E, Tpr. 80. RTL Netherlands. Symbol rate = 27.5 Msym/s, FEC rate = 3/4, CA = Irdeto/Mediaguard.

French-language digital bouquets: ASTRA 1-E, Tpr. 66 and 70. Canal Satellite. Symbol rate = 27.5 Msym/s, FEC rate = 3/4, CA = Mediaguard and Viaccess.

ASTRA 1-E, Tpr. 68 and 72. ASTRA 1-F, Tpr. 97. Canal Plus. Symbol rate = 27.5 Msym/s, FEC rate = 3/4, CA = Irdeto.

ASTRA 1-G, Tpr. 66, 68, 70, and 72. ASTRA 1-F, Tpr. 86 and 100. Canal Satellite (http://www.canalsatellite.fr). Symbol rate = 27.5 Msym/s, FEC rate = 3/4, CA = Mediaguard and Viaccess.

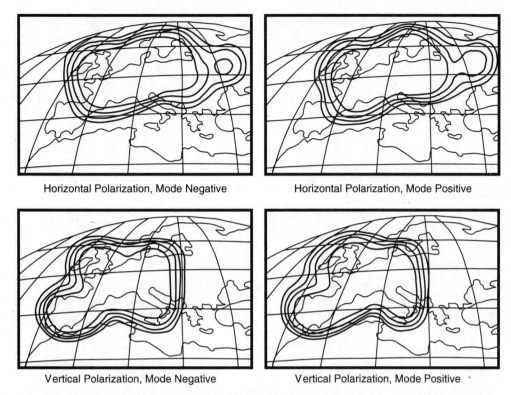

Horizontal Polarization, Mode Negative   Horizontal Polarization, Mode Positive

Vertical Polarization, Mode Negative   Vertical Polarization, Mode Positive

Contours represent 52, 50, 48 and 46 dBW or 60, 75, 90 & 120 cm antenna sizes from the innermost to the outermost contour. All transponders using vertical polarization also can be received in the Canary Islands with 120 cm dishes.

Figure 3–5 ASTRA 1-E satellite downlink coverage beams.

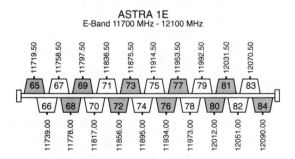

Figure 3–6 ASTRA 1-E transponder frequency plan.

CA = None, free to air and Symbol rate = 22 Msym/s, FEC rate = 5/6, CA = Mediaguard and Cryptoworks.

German-language digital bouquets: ASTRA 1-E, Tpr. 65, 69, and 81. DF-1. Symbol rate = 27.5 Msym/s, FEC rate = 3/4, CA = Irdeto.

ASTRA 1-E, Tpr. 67. Premiere. Symbol rate = 27.5 Msym/s, FEC rate = 3/4, CA = Irdeto.

ASTRA 1-E, Tpr. 82. Pro 7. Symbol rate = 27.5 Msym/s, FEC rate = 3/4, CA = Cryptoworks.

ASTRA 1-F, Tpr. 81, 83, 84, 87, and 91. Deutsche Fernsehen (http://www.dfs.de). Symbol rate = 27.5 Msym/s, FEC = 3/4, conditional access (CA) = Irdeto.

English-language digital bouquet: ASTRA 1-F, Tpr. 88. Turner Network Television (http://www.tnt-tv.com). Symbol rate = 27.5 Msym/s, FEC rate = 3/4,

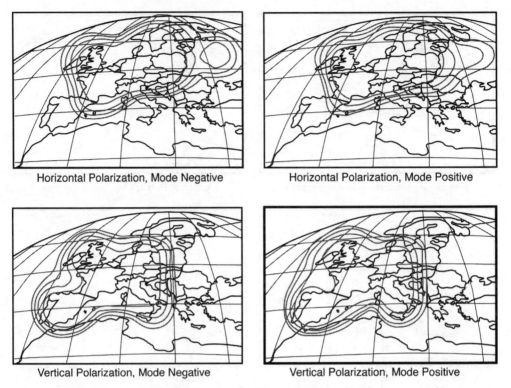

| Horizontal Polarization, Mode Negative | Horizontal Polarization, Mode Positive |
| Vertical Polarization, Mode Negative | Vertical Polarization, Mode Positive |

Contours represent 52, 50, 48 and 46 dBW or 60, 75, 90 & 120 cm antenna sizes from the innermost to the outermost contour. All transponders using vertical polarization also can be received in the Canary Islands with 120 cm dishes.

Figure 3–7 ASTRA 1-F satellite downlink coverage maps.

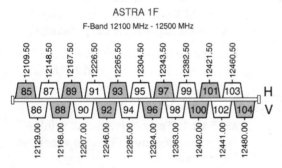

Figure 3–8 ASTRA 1-F satellite transponder frequency plan.

ASTRA 1-F, Tpr. 88. Sat 1, Kabel 1, and Deutsches Sports Fernsehen. Symbol rate = 27.5 Msym/s, FEC rate = 3/4, CA = None, free to air.

ASTRA 1-F, Tpr. 89. RTL (http://www.rtl.de). Symbol rate = 27.5 Msym/s, FEC rate = 3/4, CA = Cryptoworks.

ASTRA 1-G, Tpr. 98. Deutsche Welle (http://www.dwtv.de). Symbol rate = 22 Msym/s, FEC rate = 5/6, CA = None, free to air.

ASTRA 1-G, Tpr. 105 and 109. Canal Plus Netherlands. Symbol rate = 22 Msym/s, FEC rate = 5/6, CA = Irdeto and Mediaguard.

ASTRA 1-G, Tpr. 111 and 119. ARD (http://www.ard.de). Symbol rate = 22 Msym/s, FEC rate = 5/6, CA = None, free to air.

ASTRA 1-G, Tpr. 115. ZDF (http://www.zdf.de). Symbol rate =

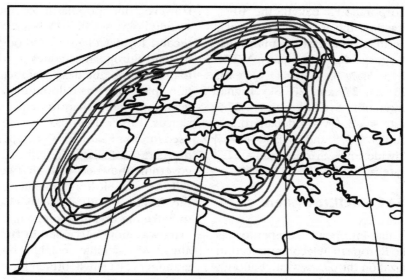

Contours represent 52, 50, 48 and 46 dBW or 60, 75, 90 & 120 cm antenna sizes from the innermost to the outermost contour. All transponders using vertical polarization also can be received in the Canary Islands with 120 cm dishes.

Figure 3–9 ASTRA 1-G satellite downlink coverage map.

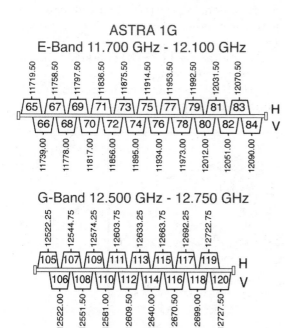

Figure 3–10 ASTRA 1-G satellite and transponder frequency plan.

22 Msym/s, FEC rate = 5/6, CA = None, free to air.

ASTRA 1-G, Tpr. 117. Austrian bouquet. Symbol rate = 27.5 Msym/s, FEC rate = 3/4, CA = None for free to air services, Cryptoworks for pay TV services.

Polish-language digital bouquet: ASTRA 1-E, Tpr. 79. ASTRA 1-F, Tpr. 99. Wizja TV. Symbol rate = 27.5 Msym/s , FEC rate = 3/4, CA = Cryptoworks.

Spanish-language digital bouquets: ASTRA 1-E, Tpr. 74, 76, and 78. ASTRA 1-F, Tpr. 92, 93, and 94. Canal Plus Spain. Symbol rate = 27.5 Msym/s, FEC rate = 3/4, CA = Mediaguard.

ASTRA 1-G, Tpr. 106 and 110. Canal Satellite Digital (http://www.csatelite.es). Symbol rate = 27.5 Msym/s, FEC rate = 3/4, CA = Mediaguard.

The Astra 2-A satellite at 28.2 degrees east longitude serves as the United Kingdom's new DTH platform for delivery of digital bouquets from News Corporation's BSkyB, the BBC, Turner Broadcasting, the Discovery Channel, and various German broadcast organizations.

English-language digital bouquets: ASTRA 2-A, Tpr. 1, 2, 3, 5, 7, 8, 9, 10, 11, 12, 18, 19, 20, 22, 23, 24, 27, and 28. Sky Digital (http://www.sky.co.uk). Symbol rate = 27.5 Msym/s, FEC rate = 2/3, CA = Videoguard.

ASTRA 2A, Tpr. 1, 3, 5, and 23. BBC (http://www.bbcworldwide.com). Symbol rate = 27.5 Msym/s, FEC rate = 2/3.

The EUTELSAT Satellite System

Astra's main pan-European competitor is EUTELSAT: The European Telecommunication Satellite Organization (http://www.eutelsat.org). EUTELSAT is a multinational satellite cooperative with its headquarters in Paris, France. Members of the European Conference of Postal and Telecommunications Administrations (CEPT) established EUTELSAT in 1977.

Four EUTELSAT I series satellites were launched during the 1980s. Although three of these satellites were still in orbit at the time of writing, these first-generation satellites perform secondary roles as an emergency backup for full-time voice and data services. (See Figures 3–11 and 3–12.)

From the beginning, EUTELSAT satellites have carried a large number of TV services.

However, the popularity of these satellites among home satellite TV households remains second to that of Astra. Although EUTELSAT has had European satellites in orbit for more than a decade, the design of their first-generation spacecraft was tailored more to handling telephony traffic than to delivering television signals. Larger antennas therefore had to be used for home satellite TV reception than what the Astra constellation required. Moreover, the government-owned PTTs that controlled EUTELSAT were slow in implementing the necessary technical and regulatory changes needed to make the organization competitive. Finally, Astra was more successful in creating the first attractive "satellite neighborhood" of programmers at a single orbital location.

During the late 1980s, EUTELSAT began designing new higher-powered satellites to accommodate an expanding universe of small-aperture satellite TV systems that serve both the commercial cable TV and home satellite TV marketplaces in Europe. At the time of writing, fifteen EUTELSAT satellites were in orbit, with four additional spacecraft scheduled for launch in the near future. The EUTELSAT II F1 through F5 satellites, which use the 10.95–11.20, 11.45–11.70, and 12.50–12.75 GHz frequency bands, can operate 16 transponders simultaneously. The HOT BIRD satellites each operate an additional 16 transponders within the 11.20–11.45

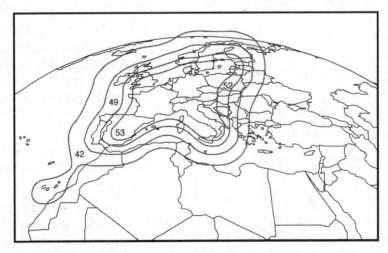

Figure 3–11 *Hot Bird 2, 3, and 4 Superbeam downlink coverage map.*

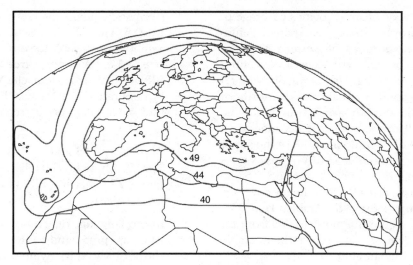

Figure 3–12 Hot Bird 2, 3, and 4 Widebeam downlink coverage map.

GHz, 11.7–12.1, 12.1–12.5, and 10.7–10.95 GHz frequency spectra (Figure 3–13).

With the launch of HOT BIRD 1 in 1995, EUTELSAT was able to emulate the Astra system by colocating satellites that operate in adjacent frequency segments of the Ku-band spectrum. With today's colocation of HOT BIRD 1, 2, 3, 4,

and 5 at 13 degrees east longitude, EUTELSAT has more than 80 transponders available from a single orbital location.

EUTELSAT primarily has assigned its digital DTH traffic to the satellite platforms located at the 13 degrees east longitude location. The HOT BIRD satellites each carry two transmit an-

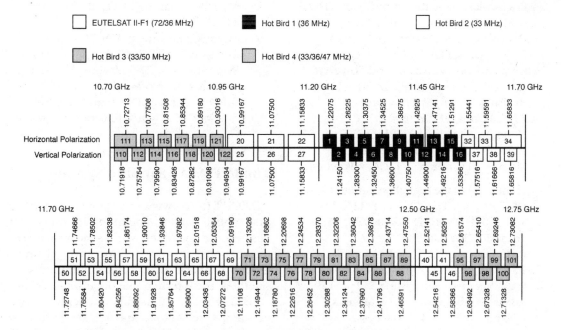

Figure 3–13 EUTELSAT II and HOT BIRD satellite transponder frequency plan.

tennas. The "wide beam" provides coverage of mainland Europe, Russia, the Arabian Peninsula, and even portions of western Asia; the "super beam" provides coverage of the central portion of the continent, including outlying regions such as southern Spain and the northern portion of the Scandinavian countries. The nominal EIRP of the wide beam's innermost contour is 49 dBW, while the EIRP of the super beam's center contour is 53 dBW. These values equate to receiving antenna apertures of 80 and 60 cm, respectively. Wide beam coverage extends through most of the Arabian Peninsula and East Asia, where antennas ranging from 1.2 to 1.8 m in diameter can receive the available digital DTH programming.

HOT BIRD Digital DTH Lineup (10.7–10.95, 11.7–12.5 GHz, Polarization: Linear)

Arabic-language digital bouquets: HOT BIRD 2, Tpr 65. Arabesque. Symbol rate = 27.5 Msym/s, FEC rate = 3/4, CA = Viacess. HOT BIRD 4, Tpr. 97. Arabsat. Symbol rate = 27.5 Msym/s, FEC = 3/4, CA = None, free to Air.

English-language digital bouquets: HOT BIRD 1, Tpr. 2 and 5. VIACOM and Discovery bouquets. Symbol rate = 27.5 Msym/s, FEC rate = 3/4, CA = Cryptoworks (VIACOM) and Powervu (Discovery). HOT BIRD 3. Tpr. 111. British Telecom. Symbol rate = 27.5 Msym/s, FEC rate = 2/3, CA = None, free to air. HOT BIRD 5, Tpr. 157. Bloomberg TV. Symbol rate = 27.5 Msym/s, FEC rate = 3/4, CA = Powervu.

French-language digital bouquets: HOT BIRD 4, Tpr. 112, 114, 116, 118, and 120. HOT BIRD 5, Tpr. 126, 158. Television Par Satellite (http://www.tps.fr). Symbol rate = 27.5 Msym/s, FEC rate = 3/4, CA = Viacess. HOT BIRD 5, Tpr. 159. ABSat. Symbol rate = 27.5 Msym/s, FEC rate = 3/4, CA = Viacess.

German-language digital bouquets: HOT BIRD 5, Tpr. 127. Deutsche Telecom. Symbol rate = 27.5 Msym/s, FEC rate = 5/6, CA = none, free to air. HOT BIRD 5, Tpr. 132. Deutsche Welle. Symbol rate = 9.1 Msym/s, FEC rate = 1/2, CA = none, free to air.

Greek-language digital bouquet: HOT BIRD 2, Tpr. 55 and HOT BIRD 3, Tpr. 73. Multichoice Hellas. Symbol rate = 27.5 Msym/s, FEC rate = 3/4, CA = Irdeto.

Italian-language digital bouquets: HOT BIRD 2, Tpr. 52 and 54. RAI (http://www.rai.it). Symbol rate = 27.5 Msym/s, FEC rate = 2/3, CA = none, free to air. HOT BIRD 2, Tpr. 57, 59, 62, 64, 66–68 and HOTBIRD 3, Tpr. 82 and 86. D Plus. Symbol rate = 27.5 Msym/s, FEC rate = 3/4, CA = Irdeto/Mediaguard.

Other digital bouquets: HOT BIRD 2, Tpr. 61. NTV Russia. Symbol rate = 20.0 Msym/s, FEC rate = 3/4, CA = Viaccess. HOT BIRD 3, Tpr. 80. TV Slovakia. Symbol rate = 27.5 Msym/s, FEC rate = 3/4, CA = Viaccess. HOT BIRD 3, Tpr. 81. TV Poland. Symbol rate = 27.5 Msym/s, FEC rate = 3/4, CA = none, free to air. HOT BIRD 3, Tpr. 88. Polsat. Symbol rate = 27.5 Msym/s, FEC rate = 3/4, CA = Irdeto/Mediaguard. HOT BIRD 5, Tpr. 90. TV Croatia. Symbol rate = 27.5 Msym/s, FEC rate = 3/4, CA = Viaccess.

The INTELSAT Satellite System

INTELSAT (http://www.intelsat.int), the world's first global satellite communications system, has become one of the world's most successful international business cooperatives. Using a global network of 22 telecommunications satellites, the international organization simultaneously relays thousands of telephone conversations, telegrams, and telexes and more than one hundred television program services between nations; in some cases, it distributes signals domestically. INTELSAT satellites also handle approximately two-thirds of the world's overseas telephone calls and most transoceanic television. In addition to the telephone and data channels made available, these satellites relay domestic TV services and provide capacity for TV news exchanges.

As many as 16 INTELSAT V, VI, VII, and VIII series satellites can be received from locations in Europe, Africa, and the Middle East

(Figures 3–14 through 3–17). These satellites are positioned in clusters located over the Atlantic Ocean Region (AOR: 53, 50, 40.5, 34.5, 31.5, 29.5, 27.5, 24.5, 21.3, 18, and 1 degrees west longitude) and the Indian Ocean Region (IOR: 33, 57, 60, 62, 64, and 66 degrees east longitude). Not all of these satellites, however, relay digital TV programming to locations within the region. INTELSAT also has a dedicated Ku-band-only satellite, INTELSAT K, which is located at 21.5 degrees west longitude over the Atlantic Ocean.

INTELSAT 605 Ku-band Spot West

EIRP contours: 47, 46, 45, 44, 43, 42 & 41 dBW

INTELSAT 605 Ku-band Spot East

EIRP contour values: 47, 46, 45 & 44 dBW

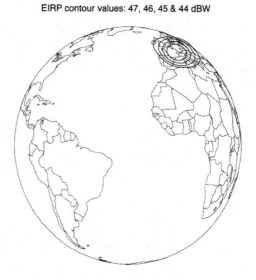

Figure 3–14 INTELSAT 605 Ku-band spot beams.

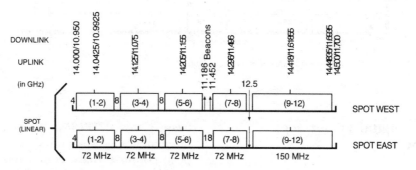

Figure 3–15 INTELSAT 605 Ku-band transponder frequency plan.

INTELSAT 707 Ku-band Spot 1

EIRP contours: 48.5, 47.5, 46.5, 45.5 & 44.5 dBW

INTELSAT 707 Ku-band Spot 2

EIRP contours: 48.5, 47.5, 46.5,
45.5, 44.5, 44.5 43.5, 42.5 & 41.5 dBW

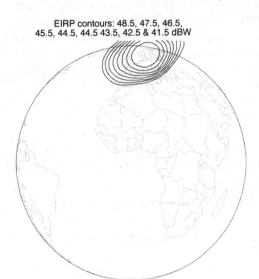

Figure 3–16 *INTELSAT 707 Ku-band spot beams.*

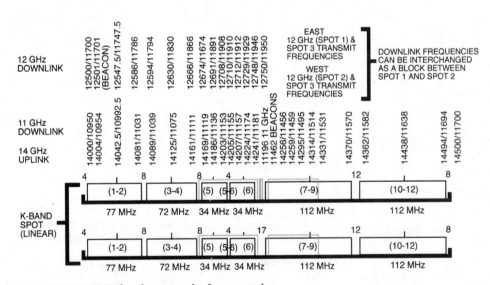

Figure 3–17 *INTELSAT 707 Ku-band transponder frequency plan.*

INTELSAT Digital TV Services for Europe

The INTELSAT VI and VIII satellites are equipped with two steerable Ku-band spot beams; the INTELSAT VII spacecraft carry three steerable Ku-band spots. Several European countries have contracted for Ku-band spot beam capacity on INTELSAT satellites located over the Atlantic and Indian Oceans. For example, British Telecom has acquired multiple transponders aboard the INTELSAT 605 satellite at 27.5 degrees west longitude for the delivery of analog and digital TV program services to cable system operators and SMATV systems. Norwegian Telecom transmits numerous digital TV services through two of the three available Ku-band spot beams on the INTELSAT 707 satellite located at 1 degree west longitude. The available services include NRK and TV Norway. The third Ku-band spot beam on this satellite, which is centered over the Middle East, is used to transmit Israeli television.

INTELSAT 703 at 57 degrees east longitude carries numerous satellite TV services for South Africa on its C-band transponders, as well as the U.S. government's Worldnet TV service. INTELSAT 703 also carries three Ku-band spot beams, one of which is boresighted onto the Middle East in order to transmit DTH TV services on behalf of Orbit Communications (28 digitally compressed video services and 12 digital audio services).

Located at 21.5 degrees west longitude, INTELSAT K is used by organizations such as n-TV and Reuters Television to digitally transmit news and sports feeds within Europe on an as-needed basis. Neither of these digital service providers, however, transmits programming that is intended for reception by the general public.

INTELSAT C-band Digital Bouquets (3.65–4.2 GHz, Polarization: Circular)

INTELSAT 605 - 27.5 West [Tpr. 21, 3.762 GHz]. Worldnet (http://www.ibb.gov/worldnet) English-language digital bouquet (Symbol rate = 8.448 Msym/s, FEC = 1/2, CA = None, free to air).

INTELSAT 602 - 62 East [Tpr. 94 and 96, 3.985 and 4.015 GHz]. South Africa Broadcast Corporation (http://www.sabc.co.za) English and Afrikaans language digital bouquet.

INTELSAT Ku-band Digital Bouquets (10.95–11.2, 11.45–11.7 GHz, Polarization = Linear)

INTELSAT 605 - 27.5 West [Tpr. 71, 10.970 GHz]. D+ (http://www.telepiu.it) Italian-language digital bouquet (Symbol rate = 27.5 Msym/s, FEC rate = 3/4, CA = None).

INTELSAT 605 - 27.5 West [Tpr. 72, 11.055 GHz]. UK TV (http://www.flextech.co.uk) English-language digital bouquet (Symbol rate = 27.5 Msym/s, FEC rate = 3/4, CA = None).

INTELSAT 705 - 18 West, [Tpr. 114 and 119; 11.500 and 11.655 GHz]. Orbit Network (http://www.orbit.net) Arabic and English language digital bouquet (Symbol rate = 24.571 Msym/s, FEC rate = 7/8, CA = Powervu).

INTELSAT 707 - 1 West, [Spot 2, H-11.014 GHz]. (Up-to-date Internet listings are at http://www.telenor.no/thor/transmissions.html)

Telenor (http://www.telenor.no) digital bouquet for Scandinavia (Symbol rate = 26.0 Msym/s, FEC rate = 3/4).

INTELSAT 707 - 1 West, [Spot 2, H-11,592 GHz]. DR1 and DR2 (Symbol rate = 17.5 Msym/s, FEC rate = 3/4).

INTELSAT 707 - 1 West, [Spot 2, H-11.174 GHz]. NRK (http://www.telenor.no) Scandinavian language digital bouquet (Symbol rate = 22.5 Msym/s, FEC rate = 3/4).

INTELSAT 707 - 1 West, [Spot 1, V-11.540 GHz]. BBC Prime (http://www.bbc.co.uk) English-language digital bouquet (Symbol rate = 26.0 Msym/s, FEC rate = 3/4).

INTELSAT 703 - 57 East, [Tpr. 71, 72 and 75, 10.992; 11.075 and 11.555 GHz]. Orbit Network (http://www.orbit.net)

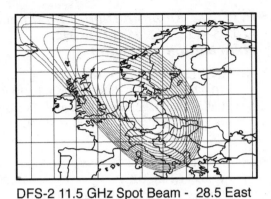

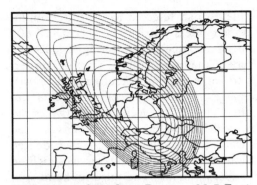

DFS-2 11.5 GHz Spot Beam - 28.5 East

Peak EIRP: 54.2 dBW. Contours
represent -1 to -15 dBW from beam center.

DFS-2 12.6 GHz Spot Beam - 28.5 East

Peak EIRP: 54.2 dBW. Contours
represent -1 to -15 dBW from beam center.

Figure 3–18 DSF Kopernikus Ku-band downlink coverage maps.

Arabic and English language digital
bouquet (Symbol rate = 24.571 Msym/s,
FEC rate = 7/8, CA = Powervu).

INTELSAT 604 - 60 East, [Tpr. 74; 11.515
GHz]. NTV (http://www.ntv.ru) Russian-
language digital bouquet (Symbol rate =
25.492 Msym/s, FEC rate = 3/4, CA =
None, free to air).

DFS Kopernikus

Located at 23.5 and 28.5 degrees east longitude, respectively, Germany's DFS Kopernikus
2 and 3 satellites are medium-powered Ku-
band telecommunications spacecraft that relay
TV programming into Western Europe within
the 11.45–11.70 GHz and 12.5–12.75 GHz
frequency spectra (Figures 3–18 and 3–19).
The available German-language TV services include arte, SAT-1, DSF Deutsche Sport Fernsehen, Premiere, Pro Sieben, and VOX. Most of
these TV services are also available from one or
more of the Astra satellite platforms. DFS satellite operator Deutsche Telekom has contracted
with SES for the relocation of all traffic from
the DFS Kopernikus satellite system to the
Astra 2B satellite following its launch to 28.2
degrees east longitude in 1999.

DSF-3 Digital Bouquets (11.45–11.7 and 12.5–12.75 GHz, Polarization: Linear)

German-language digital DTH bouquets:

DFS-3 [Tpr. A, B; 11.465 and
11.575 GHz]. DF-1 (http://www.dfs.de)
Deutsche Fernsehen 1. Symbol rate =
27.5 Msym/s, FEC rate = 3/4,
CA = Irdeto.

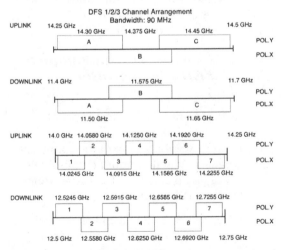

Figure 3–19 DSF Kopernikus Ku-band transponder
frequency plan.

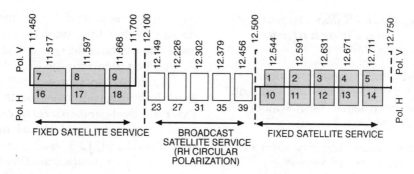

Figure 3–20 *HISPASAT Ku-band transponder frequency plan.*

DFS-3 [Tpr. A, C; 11.498 and 11.615 GHz]. ARD (http://www.ard.de). Symbol rate = 27.5 Msym/s, FEC rate = 3/4, CA = None, free to air.

DFS-3 [Tpr. K3; 12.610 GHz]. Sat 1 (http://www.sat1.de). Symbol rate = 27.5 Msym/s, FEC rate = 3/4, CA = None, free to air.

DFS-3 [Tpr. K6; 12.692 GHz]. ZDF Vision (http://www.zdf.de). Symbol rate = 27.5 Msym/s, FEC rate = 3/4, CA = None, free to air.

High-Power European BSS Satellites

Spain (Hispasat), Norway (Thor) and Sweden (Sirius) operate high-power satellites that totally conform to the WARC-77 BSS plan for ITU Region 1. Only HISPASAT 1A and 1B, however, were carrying digital DTH programming at the time of writing (Figures 3–20 and 3–21).

The Hispasat 1A and 1B satellites, which were launched in 1992 and 1993, respectively, are multiservice platforms that operate in both the BSS and FSS frequency bands. Both spacecraft are colocated at 30 degrees west longitude, just 1 degree away from Spain's WARC-assigned orbital location of 31 degrees west longitude. Hispasat's BSS spot beam coverage of mainland Spain produces a nominal EIRP of 56 dBW, which equates to a minimum antenna

size of 45 cm in diameter. The Hispasat FSS coverage beam produces a nominal EIRP of 50 dBW, which requires the use of 75-cm antennas.

Hispasat 1A and 1B BSS Payload (12.1–12.5 GHz, Polarization: Circular)

Spanish-language digital bouquets: Tpr. 23, 27, 35, and 39. Via Digital (http://www.viadigital.com). Symbol rate = 27.5 Msym/s, FEC rate = 3/4. Tpr. 31. Canal Satelite Digital (http://www.csatelite.es). Symbol rate = 27.5 Msym/s, FEC rate = 3/4.

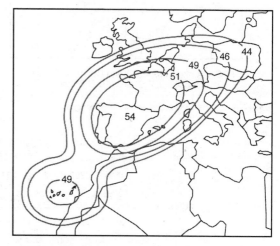

Figure 3–21 *HISPASAT Ku-band FSS satellite coverage beam.*

Hispasat 1A and 1B FSS Payload (11.45–12.1 GHz, Polarization: Linear)

Spanish-language digital bouquets: Tpr. 1, 2, 5, 11, 12, 9, and 16. Via Digital (http://www.viadigital.com). Symbol rate = 27.5 Msym/s, FEC rate = 3/4.

Tpr. 4. Retevision (http://www.rtve.es). Symbol rate = 28.116 Msym/s, FEC rate = 5/6.

PanAmSat's PAS Satellites

PanAmSat is a global satellite service provider that currently operates six PAS satellites in geostationary orbit and has plans to launch an ad-ditional four spacecraft before the end of the decade. Within the Americas, PanAmSat also owns and operates the Galaxy and SBS domestic satellite systems for the United States.

Launched on August 3, 1995, to 68.5 degrees east longitude over the Indian Ocean, PAS-4 is a dual-band spacecraft carrying 16 C-band transponders and 24 Ku-band transponders. On C-band, 12 34-watt transponders are available with a bandwidth of 54 MHz as well as 4 transponders with a bandwidth of 64 MHz. (See Figures 3–22 and 3–23.)

On Ku-band, the spacecraft features 8 54 MHz-wide transponders and 16 27 MHz-wide transponders (Figures 3–24 and 3–25). All Ku-band transponders feed into 63-watt amplifiers. PAS-4 also features five distinct regional Ku-band beams covering Southern Africa,

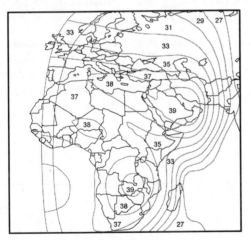

PAS-4 Africa/Europe C-band Downlink Beam
Transponders: 2-C, 6-C, 10-C & 12-C

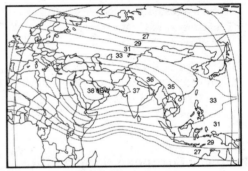

PAS-4 South Asia/Middle East
C-band Downlink Beam
Tprs: 1-C, 3-C, 5-C, 7-C, 9-C, & 11-C;
2-C, 6-C, 12-C & 16-C (switchable)

Figure 3–22 PAS-4 C-band satellite coverage beams.

DOWNLINK TRANSMIT
SOUTH ASIA/MIDDLE EAST

	3730	3790	3850	3915	3980	4040	4100	4165	
HORIZONTAL	1-C	3-C	5-C	7-C	9-C	11-C	13-C	15-C	4200
3700 VERTICAL	2-C	4-C	6-C	8-C	10-C	12-C	14-C	16-C	

AFRICA OR SOUTH ASIA/ME   ASIA   AFRICA OR SOUTH ASIA/ME   ASIA   AFRICA   ASIA OR SOUTH ASIA/ME   AFRICA   ASIA OR SOUTH ASIA/ME

├ 54 ┤ 6   ├ 64 ┤   ├ 54 ┤
3│   │   │   3│

Figure 3–23 PAS-4 C-band transponder frequency plan.

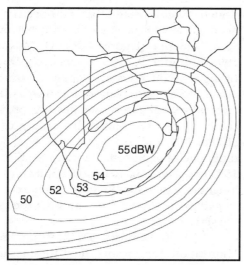

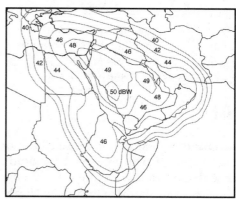

PAS-4 Arabian Ku-band Downlink Beam
Transponders: 17-K, 19-K, 21-K & 23-K

PAS-4 S. Africa Ku-band Downlink Beam
Transponders: 1-K, 2-K, 3-K & 4-K

Figure 3–24 *PAS-4 Ku-band satellite coverage beams.*

India, the Persian Gulf states, Eastern Europe, and Northeast Asia, respectively.

Within ITU Region 1, PAS-4 transmits two digital DTH bouquets on behalf of Multi-Choice Africa, a Ku-band spot beam service targeting the Republic of South Africa and a C-band digital bouquet covering all of Africa and the Arabian Peninsula. Also available: a free-to-air C-band bouquet from China Central Television (CCTV) targeting the Middle East and a Ku-band bouquet from Showtime that uses the PAS-4 satellite's Middle East Ku-band spot beam.

PAS-4 C-band Transponders (Polarization: Linear)

CCTV (http://www.cctv.com) Chinese-language digital DTH bouquet (Symbol rate = 19.850 Msym/s, FEC rate = 3/4, CA = Powervu).

MultiChoice (http://www.allicat.mck.co.za) multilingual digital DTH bouquet (Symbol rate = 20.6Msym/s, FEC rate = 3/4, CA = Irdeto).

HORIZONTAL/SOUTHERN AFRICA

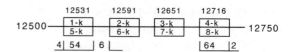

VERTICAL/SOUTHERN AFRICA OR EUROPE

HORIZONTAL/MIDDLE EAST OR EUROPE

VERTICAL/MIDDLE EAST OR EUROPE

Figure 3–25 *PAS-4 Ku-band transponder frequency plan.*

PAS-4 Ku-band Transponders (Polarization: Linear)

MultiChoice (http://www.allicat.mck.co.za) multilingual digital DTH bouquet (Symbol rate = 21.85 Msym/s, FEC rate = 5/6, CA = Irdeto).

Showtime
(http://www.showtimearabia.com)
English-language digital DTH bouquet
(Symbol rate = 19.368 Msym/s, FEC
rate = 7/8, CA = Irdeto).

Arabsat II

The dual-band Arabsat 2A and 2B satellites located at 26 and 30.5 degrees east longitude
feature Ku-band beams covering a wide zone
that stretches from Southern Europe to Northern Africa and the Middle East. Each spacecraft
carries a total of 12 Ku-band transponders operating within the 12.5–12.75 GHz frequency
range. Digital DTH programmer 1stNet currently uses Ku-band capacity aboard the Arabsat 2A satellite to relay its digital DTH bouquet
to subscribers throughout the region.

Arabsat 2A Digital DTH Bouquet (12.5–12.75 GHz, Polarization = Linear)

Ku-band Tpr. 3 and 5. 1stNet
(http://www.art-tv.net) Arabic and
English language bouquet (Symbol
rate = 27.5 Msym/s, FEC rate = 3/4,
CA = Irdeto).

ITU REGION 2: THE AMERICAS

U.S. DBS satellite service providers DirecTV,
EchoStar, and USSB currently operate high-
power broadcast satellite service (BSS) systems serving subscribers in the United States.
All of these systems conform to the 1983 RARC
BSS plan for the Americas. To date, only the
United States has implemented RARC-compli-
ant systems in the Americas. Beginning in
1999, however, Canada will have its own high-
power DBS satellite in orbit at 91 degrees west
longitude.

DirecTV and USSB

On December 17, 1993, Arianespace launched
the first high-power DBS satellite for the

United States from the European Space
Agency's launch facilities in French Guyana,
South America (Figure 3–26). On August 2,
1994, and June 9, 1995, respectively, DirecTV
(http://www.directv.com) deployed two additional BSS satellite platforms known as DBS-2
and DBS-3. All three satellites are presently
colocated in the immediate vicinity of 101
(100.8, 101 and 101.2) degrees west longitude, one of eight orbital locations (175, 166,
157, 148, 119, 110, 101 and 61.5 degrees west
longitude) that the RARC-83 conference assigned to the United States for broadcast satellite service operations.

In 1982, Hubbard Broadcasting, Inc., of
Minneapolis, Minnesota, filed the first application before the Federal Communications
Commission for a Direct Broadcast Satellite
(DBS) license. Parent company Hubbard
Broadcasting subsequently formed a new
company called United States Satellite Broadcasting (http://www.ussb.com) to develop an
integrated plan for the introduction of services. In 1991, USSB and DirecTV's parent
company Hughes Electronics Corporation
agreed to jointly develop a common digital
satellite system (DSS) that would use the
same satellite platform and digital compression technologies.

DirecTV currently delivers more than
175 channels of digital DTH programming to
homes and businesses with a DSS-compatible
receiving system that features a 60 cm dish.
The same DSS system can also access 26 additional channels from USSB, giving DSS system
owners access to a combined total of 210
channels of programming from a single satellite constellation.

EchoStar

In 1987, EchoStar Communications Corporation filed for a Direct Broadcast Satellite
(DBS) license with the FCC and established
the EchoStar Satellite Corporation (http://
www.dishnetwork.com) to build, launch, and
operate a new series of EchoStar DBS satel-

lites. The FCC awarded a DBS orbital slot at 119 degrees west longitude to EchoStar in 1992. On December 28, 1995, a Chinese Long March 2E rocket successfully launched the EchoStar I satellite to orbit. On March 4, 1996, EchoStar's DISH Network began broadcasting digital DTH services to customers throughout the United States (Figure 3–27).

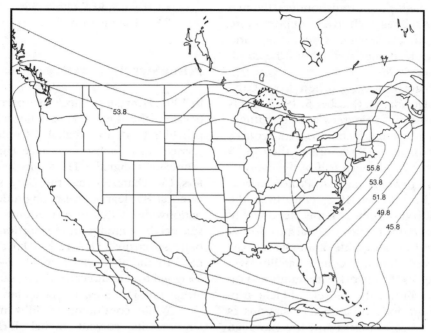

DBS-1 EIRP contours from 101 degrees west longitude

Figure 3–26 *DBS-1 satellite coverage beams.*

ECHOSTAR I EIRP contours from 119 degrees west longitude

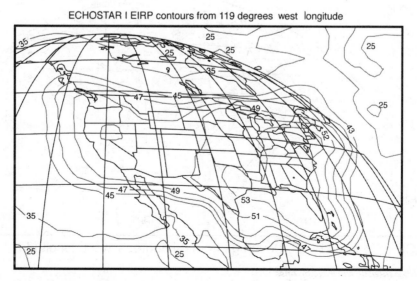

Figure 3–27 *EchoStar satellite coverage beams.*

Arianespace successfully launched the company's second DBS satellite, EchoStar II, on September 10, 1996. This second satellite is currently colocated with EchoStar I at 119 degrees west longitude. With a combined capacity of 32 high-power transponders, the two colocated satellites collectively deliver more than 180 channels of digital video, audio, and data services to homes throughout the continental United States.

EchoStar III was launched from Cape Canaveral, Florida, on October 5, 1997, and is now located at 61.5 degrees west longitude, where it is delivering niche program services and local broadcast TV stations to the top 15 TV markets in the eastern half of the country. On May 8, 1998, a Russian Proton rocket launched the EchoStar IV satellite from the Baikonur Cosmodrome in the Republic of Kazakhstan. From its new orbital position at 148 degrees west longitude off the West Coast of the United States, the new satellite will provide the DISH Network with additional capabilities, including the delivery of niche programming and local broadcast TV stations as well as coverage of Alaska and the Hawaiian Islands.

EchoStar I, II, III, and IV (12.2–12.75 GHz, Polarization = Circular).

The DISH NETWORK. Symbol rate = 20 Msym/s, FEC rate = 3/4, CA = Nagravision.

PrimeStar

In 1994, PrimeStar (http://www.primestar.com) became the first U.S. satellite TV service provider to deliver digital TV entertainment services to its subscribers (see Figure 3–28). PrimeStar's digital DTH service currently offers 160 channels that the GE-2 satellite located at 85 degrees west longitude transmits nationwide. GE-2 is a medium-power FSS spacecraft operating in FSS portion of the Ku-band spectrum (11.7–12.2 GHz) that is adjacent to the BSS spectrum (12.2–12.7 GHz) assigned to U.S. direct broadcast satellite operators DirecTV, USSB, and EchoStar.

While continuing to fully market and support its current digital DTH service, PrimeStar also plans to inaugurate a broadcast

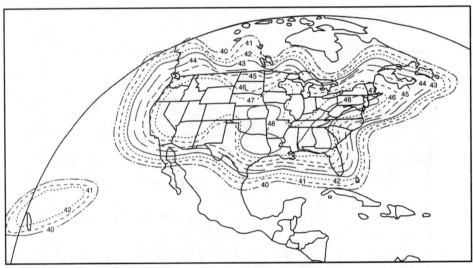

GE-2 Ku-band Beam Coverage from 85 degrees west longitude

Figure 3–28 GE-2 satellite coverage beams.

satellite service using the high-power Tempo-1 satellite located at 119 degrees west longitude using 11 FCC-licensed transponders. This high-power service is expected to deliver a variety of cable, movie, and sports channels that will be marketed in suburban and urban areas of the country.

PrimeStar also has obtained a second BSS orbital location and frequencies at 110 degrees west longitude that formerly belonged to MCI and News Corporation, together with two high-power BSS satellites that are currently under construction. (At the time of writing, PrimeStar was in the process of concluding an agreement with DirecTV that would transfer its customers and space-related assets to DirecTV.)

Canada's Nimiq DBS Satellite

Telesat Canada has contracted with Lockheed Martin for the construction of Canada's first direct broadcast satellite (DBS). The new spacecraft has been named Nimiq, an Inuit word for any object or force that unites things or binds them together. Nimiq will carry 32 high-power transponders operating in the 12.2–12.7 GHz BSS frequency band. Nimiq is scheduled for launch in 1999 to 91 degrees west longitude, one of six BSS orbital locations (138, 129, 91, 82, 72, and 70.5 degrees west longitude) assigned to Canada under the RARC-83 plan.

Two major Canadian digital DTH operators already have announced plans to migrate to the new Canadian DBS platform. ExpressVu (http://www.expressvu.com) is one of Canada's two direct-to-home television services. The other provider is Electronic Digital Delivery (EDD), which intends to use the Canadian DBS satellite as an electronic "video store in the sky" that will allow viewers to order and download movies and other types of programming.

Until Nimiq is successfully launched, Telesat Canada (http://www.telesat.ca) will continue to deliver digital DTH services to Canadian households from its fleet of Anik E FSS satellites located at 107.3 and 111.1 degrees west longitude. U.S. DBS operators are prohibited from marketing DTH services in Canada under Canadian governmental regulations.

ExpressVu

ExpressVu Inc. was the first corporation in Canada to receive a license from the Canadian government to provide a digital DTH service on a nationwide basis. Established in December 1994, ExpressVu is owned by Bell Canada Enterprises Inc., Canadian Satellite Communications Inc., and Western International Communications Ltd. EchoStar Corporation has licensed ExpressVu to use the U.S. DBS operator's DISH Network digital DTH technology and brand name in Canada.

ExpressVu currently uses 14 of Anik E2's medium-power Ku-band (11.7–12.2 GHz) transponders at the orbital location of 107.3 degrees west longitude. This capacity currently transmits 100 digital video and audio channels to subscribers throughout the nation. The transition in 1999 to the high-power Nimiq satellite will permit ExpressVu to transmit up to 180 digital DTH services through seventeen of the new satellite's 32 BSS transponders.

Star Choice

Digital DTH bouquet operator Star Choice (http://www.starchoice.com) uses medium-power Ku-band capacity on Anik E2 to deliver a subscriber-customized program package that meets individual interests (Figure 3–29). Star Choice currently offers 126 video and audio channels; 107 of these are available to subscribers in the Eastern provinces, including a diverse offering of French-language programming, and 95 are available to subscribers in Canada's Western provinces.

Anik E2 Ku-band Beam Coverage from 117.3 degrees west longitude

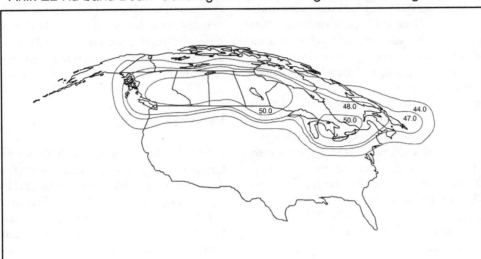

Figure 3–29 Anik E2 Ku-band satellite coverage beams.

Anik E2 (11.7–12.2 GHz, Polarization: Linear)

Tpr. 1–3, 11–14, 22–26, 31, 32. ExpressVu digital DTH bouquet. Symbol rate = 19.638, FEC rate = 3/4, CA = Nagravision.

Tpr. 4–10, 16–21, 27–29. Star Choice digital DTH bouquet. Symbol rate = 19.510, FEC rate = 3/4, CA = DigiCipher II.

C-Band Digital DTH Services

At the time of writing, there were 1.8 million households in North America equipped with C-band satellite TV receiving systems. Manufacturer General Instrument now offers a multimedia integrated receiver/decoder that can receive more than 500 digitally compressed video and audio services from North American C-band satellites. Called the 4DTV, this versatile IRD can receive analog as well as digital TV and radio transmissions. The built-in

decoder is compatible with all TV programmers using the VideoCipher RS (analog) and DigiCipher II (digital) encryption systems. The IRD also is equipped with an actuator controller that automatically steers the C-band dish to point at any North American satellite carrying digital or analog TV and radio services. Complete information concerning the available channels and satellite program lineups can be found at the General Instrument web site at http://www.4dtv.com.

Digital DTH Services in Latin America

Galaxy Latin America

In 1995, Galaxy Latin America (http://www.directvnet.com) was created to bring DirecTV digital DTH service to Latin America and the Caribbean. Galaxy Latin America, LLC ("GLA") is a multinational company consisting of DIRECTV Latin America, Inc., a Hughes Electronics company; Venezuela's Cisneros Group of Companies; Brazil's Televisao Abril; and Mexico's MVS Multivision. Based in San Jose, Costa

Rica, the company also has offices in Buenos Aires, Argentina; Caracas, Venezuela; Fort Lauderdale, USA; Mexico City, Mexico; and São Paulo, Brazil. Through local alliances in each Latin American country, GLA launched its DirecTV service in 1996. The DirecTV service is currently the leading direct-to-home television service in the 13 markets where it is available: Argentina, Barbados, Brazil, Chile, Colombia, Costa Rica, Ecuador, Guatemala, Mexico, Panama, St. Lucia, Trinidad and Tobago, and Venezuela.

On December 8, 1997, Galaxy VIII-i was successfully launched to an orbital assignment of 95 degrees west longitude. The new satellite, which is exclusively dedicated to providing DirecTV services to Latin America, carries 32 Ku-band transponders. The satellite's 115-watt transponders beam their signals to all of Latin America and the Caribbean. Sixteen transponders, which offer programming in Portuguese and Spanish, cover Brazil and the Southern Cone. The remaining 16 transponders broadcast to the rest of Latin America, the Caribbean, and the Southern Cone with programming predominantly in Spanish.

Subscribers throughout the region can receive signals from the Galaxy VIII-i satellite using antenna ranging from 60 cm to 1.1 m in diameter. Services are broadcast from five program centers located in Long Beach, California; Mexico City, Mexico; Caracas, Venezuela; São Paulo, Brazil; and Buenos Aires, Argentina.

Sky TV Latin America

Sky TV Latin America is a partnership of News Corporation, Grupo Televisa, Globo and Tele-Communications International, Inc. The organization currently uses capacity on four satellites (PAS-5, PAS-6, PAS-6B, and Solidaridad 2) to broadcast digital DTH services to several Latin American countries, including Brazil and Mexico.

Television Directa al Hogar (TDH)

Argentine-based Television Directa al Hogar (TV Direct To Home) transmits digital DTH services into Argentina using spot beam capacity on the Nahuel-1 Ku-band satellite located at 71.8 degrees west longitude. Most subscribers are located in rural areas of the country where access to cable TV is limited or unavailable.

ITU REGION 3: ASIA, AUSTRALIA, AND THE PACIFIC RIM

AsiaSat 2

Launched in late 1996 to 100.5 degrees east longitude, AsiaSat 2 is a dual-band communications satellite with 24 C-band transponders and 9 Ku-band transponders (Figures 3–30 and 3–31). The spacecraft's vast C-band coverage beam stretches across an area ranging from Eastern Europe and Turkey to eastern Asia, Australia, and New Zealand. Germany's Deutsche Welle has inaugurated a free-to-air digital DTH bouquet using one of AsiaSat 2's C-band transponders. Program services within this bouquet include five European public TV services—Deutsche Welle, TV5 France, RAI Italy, MCM, and TVE Spain—along with ten digital radio services.

AsiaSat 2 also carries a number of other digital TV services, many of which are currently transmitted in the clear, including regional Chinese broadcast TV networks, TV Laos, and EMTV from Papua New Guinea. Although home dish owners throughout the AsiaSat 2 footprint coverage area currently are receiving these digital TV channels, the primary purpose of the program originators is to distribute their services to remote cable and broadcast TV stations rather than to a DTH audience. STAR TV and Eastern TV Taiwan also use AsiaSat 2 to transmit digital bouquets to affiliated cable TV systems, while APTV, WTN and Reuters use the satellite to distribute news feeds to broadcasters. None of these digital services are currently available to digital DTH subscribers.

AsiaSat 2 (3.65–4.2 GHz, Polarization: Linear).

Tpr. 10B, Horizontal. The European Bouquet multilingual digital DTH

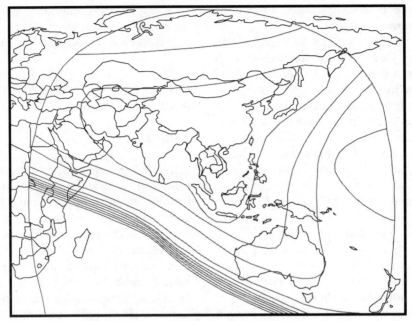

EIRP contours: 39, 37, 35, 33, 32, 31, 30, 29, 28 & 27

Figure 3-30 AsiaSat 2 C-band coverage beam.

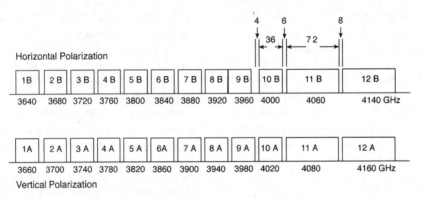

Figure 3-31 AsiaSat 2 (100.5° E) C-band transponder frequency (MHz) plan.

service. Symbol rate = 28.125 Msym/s, FEC rate = 3/4, CA = None, free to air.

Tprs. 3B, 5B, 6B, Horizontal and 5A, Vertical. Chinese provincial TV stations from Hubei, Hunan, Guandong, Liaoning, Jianxi, Fujian, Quinghai, Henan, and Guanxi. Symbol rate = 6.386 Msym/s, FEC rate = 3/4, CA = None.

Tpr. 3A, Vertical. Myanmar TV (Burma). Symbol rate = 5.08 Msym/s, FEC rate = 7/8, CA = None, free to air. Fashion TV. Symbol rate = 2.533 Msym/s, FEC rate = 3/4, CA = None, free to air.

Tpr. K2 (12.2435 GHz, horizontal). Eastern TV Taiwan Chinese-language digital DTH service. Symbol rate = 17.3617 Msym/s, FEC rate = 3/4, CA = Powervu.

Tprs 2A and 7A, Vertical. STAR TV (2) multilingual digital DTH bouquets. Symbol rate = 26.850 Msym/s, FEC rate = 7/8, CA = NDS and 28.100 Msym/s, FEC = 3/4, CA = NDS.

In 1999, Asia Satellite Telecommuncations Co., Ltd., of Hong Kong launched a new dual-band AsiaSat 3S satellite to 105.5 degrees east longitude to replace the existing AsiaSat 1 spacecraft. AsiaSat 3S features a C-band "wide beam" similar to the one carried by AsiaSat 2. Three high-powered Ku-band beams also will be available: a northern beam covering China, Japan, Korea, Mongolia, and Taiwan; a southern beam covering Afghanistan, the Indian subcontinent, Iran, Iraq, Pakistan, and Turkey; and a fully steerable spot beam. AsiaSat 3 will carry a total of 28 C-band transponders connected to the spacecraft's C-band wide beam and 16 Ku-band transponders, which can be switched between the available Ku-band coverage beams. The new spacecraft will serve as a digital DTH platform for STAR TV as well as other bouquets serving the region.

Thaicom 3

On April 16, 1997, Arianespace launched Thaicom 3, the third in a series of dual-band communications satellites for the Kingdom of Thailand. Thai satellite operator Shinawatra Satellite Public Co. Ltd. (http://www.thaicom.net) is using the new spacecraft to deliver digital DTH services, both within Thailand itself and across a vast region stretching from central Europe and Africa to eastern Asia and Australia.

Built by French manufacturer Aerospatiale, Thaicom 3 is a Spacebus 3000A three-axis stabilized satellite equipped with 25 active C-band transponders, 7 of which transmit in a semiglobal beam that spans virtually all the inhabited land masses visible from the satellite's orbital assignment at 78.5 degrees east longitude (Figures 3–32 and 3–33). The remaining 18 C-band transponders connect to a regional Asian beam that encompasses India, southern China, and Southeast Asia. The 14 Ku-band transponders on Thaicom 3 are divided between a high-powered spot beam with a nominal EIRP of 55 dBW centered over Thailand

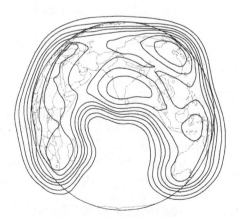

Thaicom 3 Semi-global
C-band Downlink Beam
EIRP contours: 36, 35.5, 35, 34.5,
34, 33.5, 33, 32.5, 32 dBW

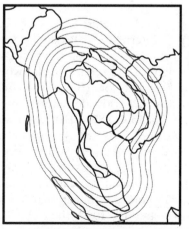

Thaicom 3 Thailand
Ku-band Downlink Beam
EIRP contours: 55, 54, 53, 52, 51,
50, 48, 46 dBW

Figure 3–32 Thaicom 3 satellite coverage beams.

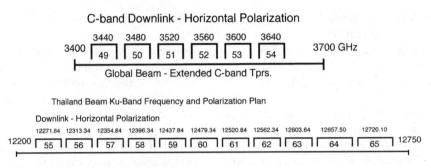

Figure 3–33 Thaicom 3 transponder frequency plans.

and a steerable spot beam that delivers a nominal EIRP of 50.1 dBW.

Twelve C-band transponders on Thaicom 3, six in the semiglobal beam and six in the East Asia regional beam, operate in the extended C-band frequency range of 3.4–3.7 GHz. This expands the number of active C-band transponders available at 78.5 degrees east longitude from 20 to 35 (25 on Thaicom 3 and 10 on Thaicom 2). The number of Ku-band transponders available at 78.5 degrees east longitude also has increased from 4 to 14 transponders, 7 of which can been connected to the steerable Ku-band spot beam that currently is focused onto the Indian subcontinent.

In August 1995, Satellite Service Co. Ltd., a subsidiary of International Broadcast Company (IBC), began transmitting the first DVB-compatible digital DTH service in the world. This digital bouquet for Thailand includes CNN International, Worldwide Entertainment, Thai Favorite Channel, ESPN, HBO Asia, Kids and Teens Channel, Kids and Special Events, and the Discovery Channel. Subscribers in Thailand also have free access to Bangkok TV channels 3, 5, 7, 9, and 11, as well as UHF broadcaster ITV and an educational TV service. On November 10, 1997, the Royal Thai Army Radio and Television began using extended C-band (3.4–3.7 GHz) transponders on Thaicom 3 to relay TV 5 programs to viewers covering Asia, Africa, Eastern Europe, and Western Australia.

Thaicom 3 C-band digital DTH (3.4–3.7 GHz, Polarization: Linear)

Tpr. 51, Horizontal. Thai TV5 Global Network (http://www.tv5.co.th) Thai-language digital DTH bouquet. Symbol rate = 26.662, FEC rate = 2/3, CA = clear.

Thaicom 3 Ku-band digital DTH (12.5–12.75 GHz, Polarization: Linear)

Tpr. 61, and 63; 12.515 and 12.600 GHz. UBC (http://www.ubctv.com) Thai-language digital DTH bouquet. Symbol rate = 25.776, FEC rate = 2/3, CA = Irdeto.

Measat 1 and 2

Licensed under the Malaysian Broadcasting Act of 1988, MEASAT Broadcast Network Systems Sdn Bhd holds the exclusive right to provide satellite broadcast services in Malaysia. Their digital bouquet is marketed under the brand name ASTRO: the All Asia Television and Radio Company.

ASTRO currently delivers 20 television and 10 radio channels to Malaysian subscribers via Ku-band capacity on the Measat 1 satellite located at 91.5 degrees east longitude. What's more, Measat 1 also is providing demonstration digital DTH services in Taiwan and the Philippines using Ku-band capacity on

Measat 1 and 2, respectively. Measat hopes to obtain licensing agreements for these two countries in the near future.

Launched on December 1, 1996, Measat 1 has the ability to transmit into a medium-powered C-band coverage beam that stretches from coastal China to Indonesia and from Burma to the Philippines. The satellite carries a total of 12 active C-band transponders, each with a bandwidth of 36 MHz. Measat 1 also carries three high power (112-watt) Ku-band spot beams centered over Malaysia, India and the Philippines (Figures 3–34 and 3–35).

Launched on November 13, 1996, Measat 2 carries six C band transponders with a bandwidth of 72 MHz and nine Ku band transponders with a bandwidth of 50 MHz. On Ku-band, Measat 2 has the ability to duplicate the coverage of Measat 1 as well as switch selected transponders to additional spot beams covering the Philippines; Taiwan; Vietnam, Laos, and Cambodia; Eastern Australia; and the Indonesian islands of Java and Sumatra.

Measat 1 Digital Bouquets (10.95–11.20 GHz and 12.25–12.5 GHz, Polarization: Linear)

Tpr. 1M, 3M, 4M, Vertical. Astro TV multilingual digital DTH bouquet. Symbol rate = 30 Msym/s, FEC rate = 3/4, CA = Cryptoworks.

Measat 2 Digital Bouquets (10.95–11.2, 11.4–11.7, 12.4–12.6 GHz, Polarization: Linear)

Tpr. 2M, 3M, 4M, Vertical. Philippines demonstration digital DTH bouquet. Symbol rate = 30 Msym/s, FEC rate = 3/4, CA = Cryptoworks.

Palapa C2 and Cakrawartha-1

Launched on May 15, 1996, to 113 degrees east longitude, Palapa C2 is a dual-band satellite equipped with 30 C-band transponders and four Ku-band transponders. Three differ-

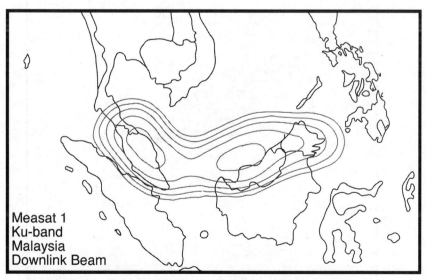

Measat 1
Ku-band
Malaysia
Downlink Beam

Effective Isotropic Radiated Power (EIRP) Contours
53, 52, 51, 50, 49, 48, 47, 46, 45, 44 dBW

Figure 3–34 Measat 1 Ku-band Malaysia coverage beam.

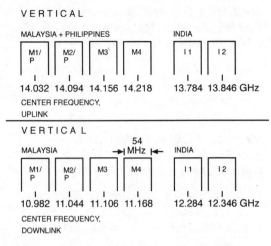

VERTICAL

MALAYSIA + PHILIPPINES   INDIA

| M1/ P | M2/ P | M3 | M4 | | I1 | I2 |

14.032 14.094 14.156 14.218   13.784 13.846 GHz

CENTER FREQUENCY,
UPLINK

VERTICAL

MALAYSIA   54 MHz   INDIA

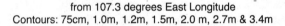

| M1/ P | M2/ P | M3 | M4 | | I1 | I2 |

10.982 11.044 11.106 11.168   12.284 12.346 GHz

CENTER FREQUENCY,
DOWNLINK

Figure 3–35 Measat 1 Ku-band transponder frequency plan.

ent C-band coverage beams are available from this satellite: a southeast Asia beam, a northern Asia beam, and an extended C-band beam that includes coverage of Australia and New Zealand.

At the time of writing, Indovision was in the process of relocating its digital DTH bouquet from Palapa C2 to the S-band Cakrawartha-1 satellite launched on November 11, 1997 (Figure 3–36). Within Indonesia, the S-band Indovision bouquet will be available to DTH receiving terminals equipped with antennas ranging from 70 cm to 1 m in diameter. In nearby countries, antennas ranging from 1 m to 2 m in diameter also will be able to receive the service.

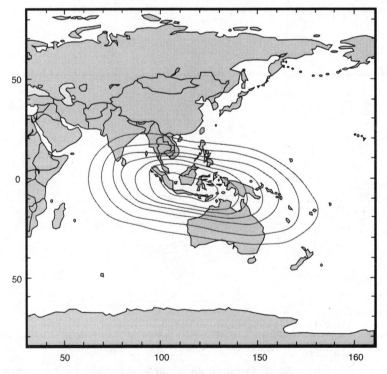

from 107.3 degrees East Longitude
Contours: 75cm, 1.0m, 1.2m, 1.5m, 2.0 m, 2.7m & 3.4m

Figure 3–36 Cakrawartha-1 S-band coverage beam from 107.3 degrees east longitude.

Optus B3

Australia's Ku-band Optus B1 and Optus B3 satellites, which transmit in the 12.25–12.75 GHz frequency range, each carry 15 Ku-band transponders and multiple spot beams, which concentrate their signals over selected portions of the Australian continent and New Zealand (Figures 3–37 and 3–38). The Australis digital DTH service introduced to Australian TV viewers in 1995 has been acquired by cable TV operator Foxtel, which is attempting to capture the former digital DTH customers of the Australis Galaxy TV bouquet. Optus Communications Pty. Ltd. (http://www.optus.net.au), operator of the Optus B satellite system, has an Optusvision digital DTH package on Optus B3 at 156 degrees east longitude that includes 16 TV channels.

In 1997, New Zealand's Sky Television began offering a single pay TV channel in an analog TV format using one of the Optus B1 transponders that connect to the New Zealand coverage beam. Sky Television intends to begin offering a digital DTH bouquet that will transmit via three New Zealand spot beam transponders on Optus B1 at 160 degrees east longitude.

JCSat 3, JCSat 4, and Superbird C

Launched on August 29, 1995, to 128 degrees east longitude, JCSat 3 is a dual-band satellite with 12 C-band transponders and 28 Ku-band transponders. JCSat 3 carries four Ku-band coverage beams for mainland Japan, northeast Asia, India, and Australia/New Zealand, as well as a C-band coverage beam that encompasses Japan, mainland China, Southeast Asia, and India.

In 1996, PerfecTV (http://www.perfectv.co.jp) launched a digital MPEG-2 bouquet

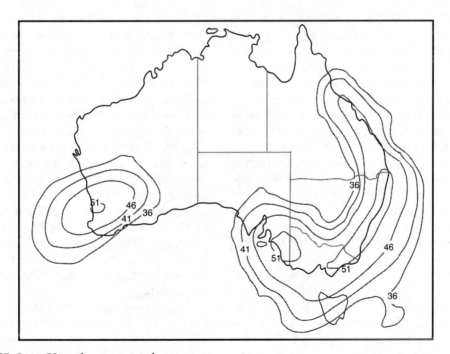

Figure 3–37 *Optus B3 southeast coverage beam.*

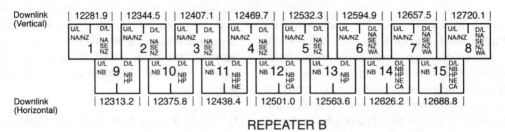

NA	National Beam, repeater A	NB	National Beam, repeater B
SE	South Eastern Australia Beam	NE	North Eastern Australia Beam
CA	Central Australia Beam	WA	Western Australian Beam
PNG	Papua New Guinea Beam	SWP	South West Pacific Beam
NZ	New Zealand Beam	HP	High Performance Beam

Figure 3–38 *Optus B3 transponder frequency plan.*

for Japanese TV viewers using this satellite's mainland Japan Ku-band beam. More than 90 TV services were part of this program package at the time of writing.

Located at 144 degrees east longitude, Superbird C is a Ku-band satellite with 24 (90-watt) Ku-band transponders, The new spacecraft will provide telecommunications services via four different Ku-band beams: Japan (including Ogasawara), Northeast Asia, Southeast Asia, and a fully steerable spot beam.

In October of 1996, Superbird operator Space Communications Corporation (SCC) established the company DirecTV Japan (http://www.directv.co.jp) in conjunction with Hughes DirecTV of the United States and major Japanese shareholders, including Culture Convenience Club, Matsushita Electric Industrial Co. Ltd., Dai Nippon Printing Company, Mitsubishi Corporation, and Mitsubishi Electric Corporation. In November of 1997, DirecTV Japan began delivering to Japanese TV viewers a digital bouquet containing more than 100 digitally compressed TV and audio services. The new service uses the East Asian beam on Superbird C. Capacity on Superbird C also provides Internet connectivity for users of the DirecPC Japan satellite delivery system.

PAS-2

PanAmSat's (http://www.panamsat.com) PAS-2 satellite was launched on July 8, 1994. PAS-2 is a dual band satellite carrying 16 C-band transponders with 34-watt power amplifiers and 16 Ku-band transponders with 63-watt power amplifiers (Figures 3–39 through 3–42). The spacecraft's C-band transponders connect to one of three Pacific Rim beams or an Oceania beam; Ku-band transponders may be connected to the China, Northeast Asia, or Australia/New Zealand beams.

PAS-2 carries numerous digital DTH services, which are available for home viewing in designated countries within the region. These include BBC World, Bloomberg Financial, Japan Entertainment Television (JET TV), CCTV 3 and CCTV 4, EWTN, The Filipino Channel, NHK World, and National Geographic. All of these services use the Powervu digital encoding system developed by Scientific Atlanta.

In 1998, PanAmSat launched a second satellite for the Pacific Ocean Region. Called PAS-8, it is located at an orbital assignment of 166 degrees east longitude. PAS-8 features 24 C-band (50-watt) and 24 Ku-band (100-watt)

PAS-2 C-band Pacific Rim Vertical #2 Downlink Beam
Transponders: 9-C & 13-C; 11-C & 15-C Switchable.

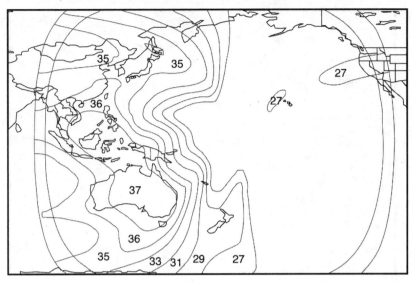

Figure 3–39 *Pas-2 C-band downlink coverage beams.*

PAS-2 C-band Pacific Rim Horizontal Downlink Beam 2
Transponders: 2-C, 6-C, 10-C & 14-C

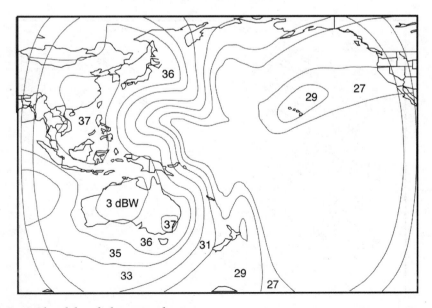

Figure 3–40 *Pas-2 C-band downlink coverage beams.*

PAS-2 C-band Pacific Rim Vertical Downlink Beam 1
Transponders: 3-C; 7-C, 11-C & 15-C switchable.

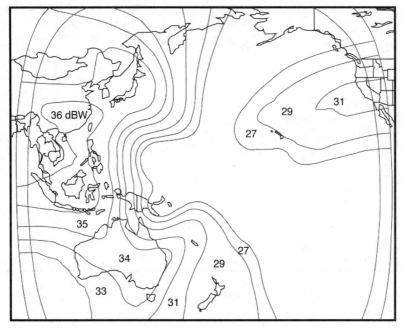

Figure 3–41 Pas-2 C-band downlink coverage beams.

PAS-2 C-band Pacific Rim Horizontal Downlink Beam 1
Transponders: 4-C, 8-C, 12-C & 16-C

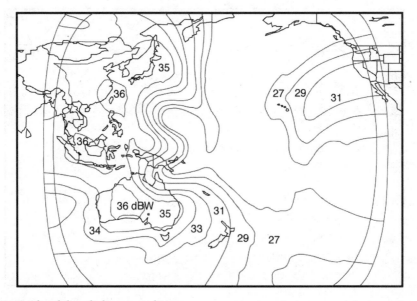

Figure 3–42 Pas-2 C-band downlink coverage beams.

transponders, each with a bandwidth of 36 MHz. The new spacecraft is expected to offer complementary services to its existing PAS-2 satellite in the Asia/Pacific region.

PAS-2 C-band Digital DTH Bouquets (3.7–4.2 GHz, Polarization: Linear)

Tpr 13-C [4,093 MHz] Vertical. National Geographic (http://www.nationalgeographic.com). Symbol rate = 29.473 Msym/s, FEC rate = 3/4, CA = n/a.

Tpr 9-C [3,962 MHz] Vertical. Japan Entertainment Television (JET). Symbol rate = 13.470 Msym/s, FEC rate = 1/2, CA = Powervu.

Tpr 1-C [3,716 MHz] Vertical. Chinese Central Television (http://www.cctv.com). Symbol rate = 19.850 Msym/s, FEC rate = 3/4, CA = None, free to air.

Tpr. 16-C [4,153 MHz] Horizontal. RAI Italy/Art Australia. Symbol rate = 5.632 Msym/s, FEC rate = 3/4, CA = Powervu.

Tpr. 2-C [3,743 MHz] Horizontal. LBC, ART, Jordan TV and CNBC. Symbol rate = 19.465 Msym/s, FEC rate = 7/8, CA = Powervu.

Tpr. 4-C [3,776 MHz] Horizontal. Discovery. Symbol rate = 19.85 Msym/s, FEC rate = 3/4, CA = Powervu.

Tpr. 8-C [3,901 MHz] Horizontal. California Bouquet (ATN, BBC World, and Bloomberg TV). Symbol rate = 30.8 Msym/s, FEC rate = 3/4, CA = Powervu.

PAS-2 Ku-band Bouquets

Tpr. 2-K [12.265 GHz], Vertical. Golden West Network (http://www.gwn.com.au) Australia. Symbol rate = 16.20 Msym/s, FEC rate = 1/2, CA = Powervu.

PAS-8 C-band Digital DTH Bouquets (3.7–4.2 GHz, Polarization: Linear).

Tpr. 12-C [3,990 MHz] Horizontal. EWTN. Symbol rate = 27.69 FEC rate = 7/8, CA = None, free to air.

Tpr. 18-C [4,060 MHz] Horizontal. NHK World (http://www.nhk.org.jp). Symbol rate = 26.470 Msym/s, FEC rate = 3/4, CA = Powervu.

INTELSAT Pacific Ocean Region (POR) Satellites

As many as eight INTELSAT VI, VII, and VIII series satellites can be received from locations in Asia and the Pacific Rim. These satellites are positioned in clusters located over the Indian Ocean Region (IOR: 57, 60, 62, 64, and 66 degrees east longitude) and the Pacific Ocean Region (POR: 174, 177, 180, and 183 degrees east longitude). Only the INTELSAT 701 satellite at 180 degrees east longitude carries digital DTH services, however. The C-band side of Intelsat 701 relays a five-channel digital DTH bouquet via an east hemispheric beam to French overseas territories that includes French-language TV programming from RFO and Canal+, and the Sports Pacific Network (SPN), which originates from the island of Nauru, via a global beam.

INTELSAT Bouquets (3.7–4.2 GHz, Polarization: Circular)

Tpr. 9 [4,095 MHz] E. Hemi beam, left-hand circular. RFO/Canal+ (http://www.rfo.fr) French-language digital DTH bouquet. Symbol rate = 27.5 Msym/s, FEC = 3/4, CA = n/a.

Tpr. 10 [4,081 MHz] Global beam, right-hand circular. Sports Pacific Network English-language sports service. Symbol rate = 4.730 Msym/s, FEC rate = 3/4, CA = None, free to air.

Chapter 4

Feedhorns and LNBs

The paraboloid is the classic shape of the antenna reflector used to receive satellite TV signals. This parabola of revolution has the property such that all incident rays arriving in parallel to the axis of symmetry are reflected to a common point or focus located to the front and center of the reflector (Figure 4–1). A perfect—that is to say, theoretical—dish would generate a sharp, well-defined focal point. In the real world, however, deviations in the accuracy of this parabolic curve, as well as the imprecise assembly of the antenna itself, often result in the generation of what might better be described as a less-defined "focal cloud." It is the task of the feedhorn to gather the signal that arrives in the vicinity of the focal point and conduct it on to the receiving system's first stage of electronic amplification: the low-noise block downconverter (LNB).

Molecular motion within all matter generates a noise background that permeates the entire electromagnetic spectrum used to propagate communication signals, including the

satellite frequency bands. The temperature of all thermal noise is expressed on the Kelvin scale, which can be related to other more familiar temperature scales such as Celsius or Fahrenheit (Figure 4–2).

Absolute zero, or 0 K, is the noise temperature at which all molecular motion stops. Even in deep space, however, the noise temperature exceeds absolute zero by several degrees. The clear sky is said to be a cold noise source because it has a noise temperature of only about 30 K, while the Earth is called a hot noise source because it has a noise temperature of about 290 K.

Although the parabolic antenna tilts up toward the "cold" sky to receive satellite signals, the feedhorn is actually pointing back toward the reflector as well as the section of the Earth behind it. It therefore is extremely important to control the way that the feedhorn "illuminates" the reflector (Figure 4–3). If the feedhorn overilluminates the dish, the feed also will see the hot noise source of the Earth

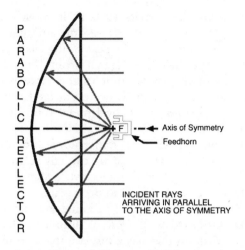

Figure 4–1 *The paraboloid is a parabola of revolution with the property of reflecting all incident rays arriving along the axis of symmetry to a common point or focus.*

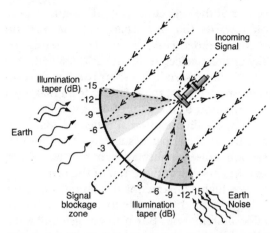

Figure 4–3 *The feedhorn illuminates the inner areas of the antenna with greater efficiency than the outer areas in order to reject noise emanating from the Earth behind the dish.*

that lies just beyond the antenna's rim. In this case, the noise temperature of the Earth will combine with the noise temperature of the satellite receiving system, thereby reducing the intensity of the desired signal. If the feedhorn underilluminates the dish, the feed will not effectively see the signal contributions that

are reflected by the outer portions of the antenna. This will reduce the amplification factor or "gain" of the dish.

The illumination taper of a simple feedhorn is a compromise between antenna gain and overillumination loss (Figure 4–4). The feedhorn receives maximum signal from the

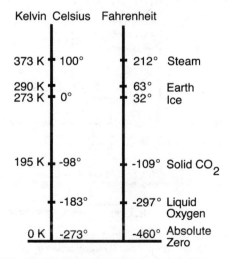

Figure 4–2 *Temperature scale conversion chart.*

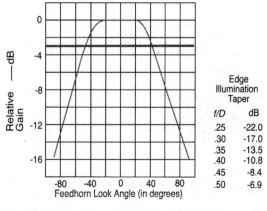

Figure 4–4 *The illumination taper is a function of the antenna's focal distance to diameter (f/D) ratio.*

center of the reflector, while the signal tapers off towards the rim. There is, however, no sharply defined cutoff point. The antenna designer must select an illumination at reflector edge that is not so low that it prohibits the outermost portion of the reflector from making a contribution to the gain of the antenna. Conversely, the illumination at reflector edge cannot be so high that a substantial amount of random noise is collected from the "hot" Earth that lies just beyond the rim of the dish.

A high gain antenna design must employ a large (–10 dB relative to center) value of edge illumination. But if low noise and sidelobes are the antenna designer's goal, a lower value of edge illumination will be required (up to –18 dB). For the small aperture paraboloid, the antenna designer may compromise by using a value of edge illumination that is in the vicinity of –14 or –15 dB, where the outer area of the paraboloid acts more as an Earth shield than as a contributor to the gain of the antenna (Figure 4–5). This is called the feedhorn illumination taper.

THE SCALAR FEEDHORN

The scalar feedhorn, which has a large circular plate with a series of three or four concentric rings attached to its surface, is most often installed on C-band receive-only antenna systems (Figure 4–6). The scalar ring plate conducts the incoming signal from the outer edges of the focal point to the large waveguide opening that is located at feed center. The waveguide is an oval metal pipe that is commonly used as a transmission line for microwave signals. The dimensions and tolerances of the waveguide are directly related to the wavelength of the microwave signals that the device is designed to propagate.

The focal length (f) to antenna diameter (D) ratio, called the f/D, is another antenna specification that has a direct impact on feedhorn performance. The distance between the scalar ring plate and the waveguide opening often is adjustable so that the installer can precisely match the feed to the manufacturer's f/D specification for the dish. This allows the scalar feedhorn to achieve optimum illumina-

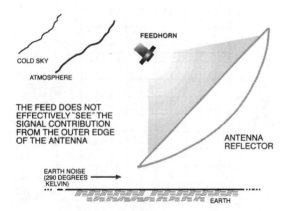

Figure 4–5 Feedhorn underillumination of the parabolic reflector.

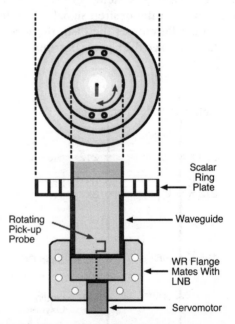

Figure 4–6 The servomotor-equipped scalar feedhorn mechanically rotates a small pickup probe to discriminate between orthogonal senses of polarization.

tion of the antenna reflector. Antenna *f/D* ratios for a prime focus parabolic antenna typically range from 0.25 to 0.45, with 0.4 the most commonly encountered.

OFFSET ANTENNA FEEDS

Virtually all of the small aperture Ku-band dishes sold today employ a design that places the focal point below the front and center of the reflector. The offset-fed antenna is oval, with a minor axis (left to right) that is narrower than its major axis (top to bottom). The offset-fed reflector only uses a small oval subsection derived from a much larger parabola of revolution. This results in a longer focal length that is typically encountered with a prime focus antenna.

The offset-fed antenna typically has a focal length to antenna diameter (*f/D*) ratio of 0.6 to 0.7. Because of its distinctive shape, the offset-fed antenna also requires a feedhorn that has been designed to illuminate the oval reflector shape. A sharp feedhorn illumination taper is not required because the feedhorn points up at the "cold" sky rather than down at the Earth. The offset-fed antenna and feedhorn most often are purchased together as a single unit.

LINEAR POLARIZATION

Geostationary communication satellites may employ linear (horizontal/vertical) or circular (right-hand/left-hand) polarization to transmit their signals back to Earth. The term polarization refers to the direction of the incoming wavefront's electric field vector. For example, a vertically polarized wave has an "E" plane that is aligned in the up–down direction and an "H" plane that is aligned in the left–right direction. Moreover, a right-hand circular wavefront has an electric field vector that rotates in a clockwise direction along its direction of propagation, while a left-hand circular wave-

front has an electric field vector that rotates in the counterclockwise direction.

Several manufacturers offer feedhorn products that can receive both the linear and circular polarization formats (Figure 4–7). In some cases a linear polarization feedhorn also can be modified to receive circular polarization signals with the addition of a rectangular insert made from a dielectric material such as Teflon.

POLARIZATION ARTICULATION

Communication satellites achieve frequency reuse by overlapping two groups of channels using orthogonal (i.e., at right angles) senses of polarization (Figure 4–8). Orthogonal does not refer only to horizontal versus vertical; it can also be applied to opposite senses of circular polarization.

A correctly adjusted linear-polarization feed will achieve peak signal acquisition in one direction of polarization, while producing a null in signal acquisition 90 degrees of rotation away. In the same way, a circular-polarization feed that is set to receive right-hand circular signals will exhibit zero response to a signal employing left-hand rotation.

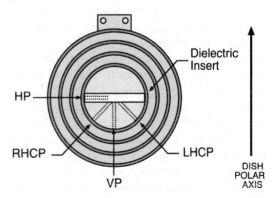

Figure 4–7 Feedhorns are available that can receive both circular and linear polarization signals.

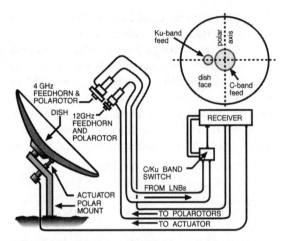

Figure 4–8 Dual-band satellite TV receiving system layout using separate C-band and Ku-band feedhorns.

Satellite operators use this orthogonal property of microwave polarization to generate a measure of isolation between two independent information streams sharing the same frequency spectrum. In this way, two signals that otherwise would interfere with each other remain separate.

An isolation of approximately 30 dB (1,000 times power ratio) between two orthogonal senses of polarization theoretically can be obtained. However, this optimum isolation value will be degraded whenever small errors in alignment or adverse weather conditions occur along the signal path. A real-world isolation figure of only 20 dB (100 times) is therefore more often the rule. As a measure of protection, satellite transponder center frequencies typically, but not always, are staggered from one set of like-polarization transponders to the other. This half-transponder offset configuration places the highest energy region of a horizontal transponder into the low-energy region of the vertical transponder set, and vice versa.

The rotating probe inside the feedhorn is rarely aligned to the vertical or horizontal plane of the site location. Rather, the terms vertical and horizontal in this instance refer to these planes from the perspective of the satellite itself, which has an orientation that is stabilized in the north-south direction in reference to the Earth's polar axis.

If the receiving site location is on the same meridian as the satellite, its horizontal and vertical planes also will match the satellite's definitions for horizontal and vertical, but elsewhere it will not. Because of the curvature of the Earth, a satellite to the east or west of the site location's meridian will appear to have its polarization axes tilted in either a clockwise or a counterclockwise direction. If the antenna reflector uses the modified polar mount, polarization tilt is automatically taken care of as the dish rotates about its polar axis.

In order to select the correct polarization, the feedhorn may incorporate a small probe that physically rotates until peak reception is obtained. The probe is rotated by means of a small servomotor that receives power from the indoor receiver or IRD. By sensing the strength of the incoming signal, some receivers can select the correct polarization setting automatically. However, most receivers are programmed either at the factory or during the installation process to recall the correct polarization format for each individual satellite stored in memory.

Many digital DTH systems are equipped with a ferromagnetic device that electronically adjusts feedhorn polarization, instantaneously and silently. This introduces a small amount of signal loss, typically 0.1–0.2 dB, which for most applications is negligible. Ferromagnetic polarization devices also have no moving parts that can cause maintenance problems in the future.

DUAL-BAND FEEDS

The dual-band feed places both the C- and Ku-band waveguide openings directly over the focus of the antenna, providing the satellite receiver with direct access to all of the TV ser-

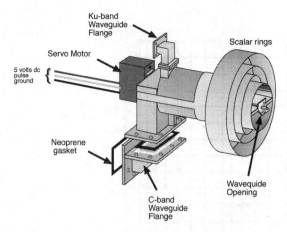

Figure 4–9 Close-up view of a dual-band feedhorn.

vices carried on dual-band satellites (Figure 4–9). The placement of both the C- and Ku-band feed openings in such close proximity, however, reduces the level of C-band satellite TV signals to less than what a good C-band-only feed can achieve. This may be an important consideration for system designers who wish to use the smallest dish possible to receive C-band satellite TV services.

THE LOW-NOISE BLOCK DOWNCONVERTER

The incoming satellite signal propagates down the waveguide of the feedhorn and exits into a rectangular chamber mounted at the front of the low-noise block downconverter (LNB), in which a tiny resonant probe is located. This pickup probe, which has a wavelength that resonates with the incoming microwave frequencies, conducts the signal onto the first stage of electronic amplification.

In addition to amplifying the incoming signal, the first stage of electronic amplification also generates thermal noise internally. The internal noise contribution of the LNB is amplified along with the incoming signal and passed on to succeeding amplifier stages.

The LNB sets the noise floor for the satellite receiving system. Today's high-performance LNB uses gallium arsenide semiconductor and high electron mobility transistor (HEMT) technologies to minimize the internal noise contribution of the LNB.

The noise performance of any C-band LNB is quantified as a noise temperature measured in kelvins (K), while Ku-band LNB noise performance is expressed as a noise figure measured in decibels (Figure 4–10). Today's C-band LNB commonly achieves a noise temperature of 40 K or less, while Ku-band noise figures of less than 1 dB/K are commonly available. In either case, the lower the noise performance rating of the LNB, the less noise introduced into the LNB by its own circuitry.

The noise performance of the LNB usually is not totally flat across the pass band of the device. There typically is a performance curve for each LNB that indicates a lower noise temperature or noise figure in some band segments over others. The manufacturer's noise performance rating actually represents the highest noise level that occurs anywhere within the pass band of the LNB.

A rectangular flange (WR-75 or WR-229) on the back of the feedhorn mates with a similar flange located at the front of the LNB (Figures 4–11 and 4–12). The installer must insert a neoprene gasket between these two flanges to prevent any moisture from entering through this seam. During assembly, this gasket must be seated correctly, as any moisture entering the waveguide will degrade signal reception and possibly damage either or both of these components. Most digital DTH systems use a product that combines the feedhorn and LNB into a single sealed unit called the low-noise feed (LNF) that eliminates this moisture entry point.

The IF output connector on the back of every LNB is another potential entrance point for moisture. The installer must seal this junction from the weather after attaching the coaxial cable to the IF output connector. The best method is to use a special waterproofing

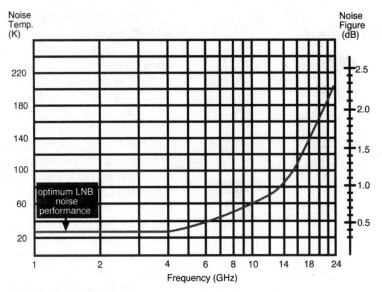

Figure 4–10 Noise temperature to noise figure conversion chart.

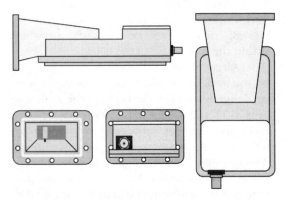

Figure 4–11 LNB side, front, back and top views.

compound such as coax-seal, which wraps tightly around the outside of the connection. Alternatively, the installer can flood the inside of the coaxial cable's F connector with a waterproofing silicon sealer. When flooding the F connector, be sure to first unplug the receiver and wait for the compound to dry before plugging the receiver back into an AC power source.

Ku-band satellites operate within distinct frequency subbands or spectra. Within any given region, for example, individual Ku-band satellites may be using the 10.95–11.7 GHz, the 11.7–12.25 GHz, and the 12.25–12.75 GHz frequency spectra. Care should be taken to ensure that each receiving system uses an LNB that matches the Ku-band frequency spectrum or spectra for the satellites that the customer desires to view.

THE LNB ISOLATOR

An isolator is a passive device that electronically separates the LNB input from the first stage of electronic amplification. An isolator performs several functions. It attenuates signals that are located outside of the band of frequencies that the LNB is designed to pass. This helps to prevent out-of-band signals, such as those produced by nearby terrestrial microwave stations, from overloading or "swamping" the first gain stage. Moreover, the

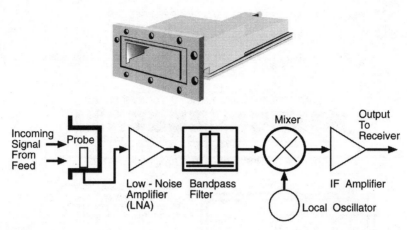

Figure 4–12 *Block diagram of an LNB.*

isolator also prevents internally generated signals, such as the local oscillator of the LNB, from exiting the amplifier, and it provides a good impedance match between the feed and initial stage of electronic amplification. Since the isolator does have some insertion loss, which adds to the amplifier's production of internal noise, not every LNB manufacturer uses one.

BLOCK DOWNCONVERSION

Every LNB contains a local oscillator (LO) that generates a microwave reference frequency. Both the incoming satellite frequency band and the LO frequency are injected into a mixer circuit, where the two signals beat or "heterodyne" to produce an intermediate frequency (IF) band that contains all of the information present in the original satellite band (Figure 4–13). The process of producing this IF signal is called block downconversion.

The cost of low-loss cables that can conduct microwave frequencies is prohibitive. The LNB must therefore downconvert the entire satellite frequency band of interest to a lower IF band that can be transferred to the indoor unit by means of a low-cost coaxial cable.

Inside the mixer circuit, the incoming satellite frequency and the LO frequency heterodyne to produce the sum (incoming satellite signal + LO) and difference (incoming satellite signal – LO) frequencies.

A C-band LNB typically uses a LO frequency of 5,150 MHz, which is higher than the frequency range of the incoming satellite band. This is called high-side frequency injection. When an LO frequency of 5,150 MHz heterodynes with the standard 3,700–4,200 MHz satellite frequency band, it produces the sum (8,850–9,350 MHz) and difference (1,450–950 MHz) frequency bands. A filter is used to reject the higher 8,850–8,350 MHz band while passing the lower 1,450–950 MHz difference frequencies onto the first IF amplifier.

From this example it can be seen that the order of the original satellite band is inverted whenever the high-side injection method is used to produce the IF band. That is, information contained in the lower portion of the satellite band (3,720 MHz) will appear in the upper portion of the 950–1,450 MHz IF band (5,150 – 3,720 = 1,430 MHz).

Several regional satellites are now transmitting digital DTH signals in what is known as the extended C-band (3,400–3,700 MHz) frequency range, as well as signals in the standard

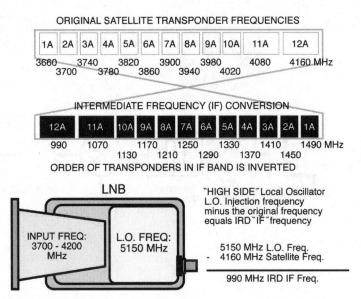

Figure 4–13 *All LNB units contain a "local" oscillator that produces a frequency that beats or "heterodynes" with the incoming satellite frequency to generate an IF band of intermediate frequencies.*

C-band (3,700–4,200 MHz) frequency spectrum. In this case, the C-band LNB local oscillator of 5,150 MHz heterodynes with the entire C-band (3,400–4,200 MHz) frequency range to produce the wide-band IF output (5,150 – 3,400–4,200 GHz = IF output of 1,750–950 MHz). A C-band LNB designed for standard C-band operations may not work well when receiving the extended C-band frequency range. The noise temperature performance of the LNB may rise dramatically when receiving extended C-band signals, and some of the lower frequency signals may not be received at all. Installers may be required to upgrade the LNB for C-band receiving systems equipped with older units. Care should be taken when ordering a replacement LNB to ensure that the new unit is capable of functioning within this extended C-band frequency range.

A Ku-band LNB typically will use an LO frequency that is lower than the frequency range of the incoming satellite band. This is called low-side frequency injection. For example, when an LO frequency of 11.3 GHz heterodynes with the 12.25–12.75 GHz satellite frequency band, it produces the sum (23.55–24.05 GHz) and difference (950–1,450 MHz) frequency bands. The order of the original satellite band is not inverted whenever the low-side injection method is used to produce the IF band (Figure 4–14).

During the initial setup of a digital IRD, the installer may need to enter the IF frequency for a given digital bouquet's default transponder. If only the original satellite frequency of the default transponder is known, the installer will need to calculate the corresponding IF frequency for that transponder. To do this, the installer will need to know the LO frequency of the LNB. This usually is printed on the label attached to each unit. A digital DTH bouquet with a Ku-band satellite frequency of 11.720 GHz and a Ku-band LNB with a local oscillator frequency of 10.75 GHz (11.720 – 10.75 = 0.97 GHz) would produce an IF center frequency of 970 MHz.

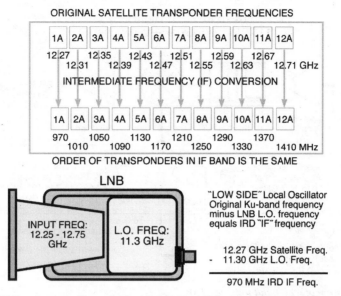

Figure 4–14 *C-band LNB units typically use high-side injection of the LO, while Ku-band units employ the low-side injection method.*

LOCAL OSCILLATOR STABILITY

Wide variations in outdoor temperature can affect the frequency stability of the LNB. Most units employ a dielectric material at the heart of the oscillator that produces the LO frequency, which is referred to as a dielectric resonant oscillator (DRO).

The dielectric slab, which is shaped to resonate at a specific frequency, usually is seated in a metal chamber. Wide temperature variations over the course of each day cause the chamber holding the dielectric to expand or contract. This causes the LO to drift by as much as 1 MHz in either direction from its nominal center frequency, which is well within the operating range of set-top boxes designed for analog TV applications. It usually will be possible to receive a digital DTH signal transmitted in an MCPC format when using an LNB with these frequency stability characteristics. However, the stability of the DRO may be insufficient for reception of some dig-

ital satellite TV transmissions using the single channel per carrier (SCPC) format. In this case, the system designer must use a phase-locked LNB that employs a highly stable crystal oscillator reference. The phase-locked LNB typically has a stability rating of [±]15 kHz, as opposed to an LNB with a dielectric resonant oscillator, which may have a stability rating of [±]500 kHz.

UNIVERSAL LNB

A wide band product called a "universal" Ku-band LNB is available that can switch electronically between the 10.7–11.7 and 11.7–12.75 GHz frequency spectra to provide complete coverage of the entire Ku-band frequency range (Figure 4–15). The receiver or IRD sends a switching voltage (13 or 17 volts DC) to the LNB that automatically changes the LNB input frequency range to the desired frequency spectrum (10.70–11.75 GHz or 11.7–12.75 GHz).

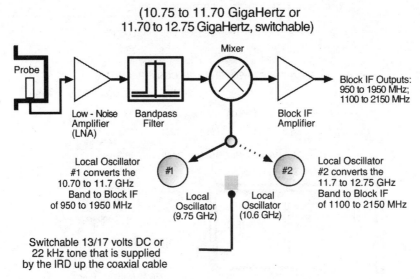

Figure 4–15 *Block diagram of a Ku-band universal LNB.*

Keep in mind, however, that any universal LNB with an IF output frequency range of 950–2,050 MHz can only be used effectively with a receiver or IRD that also has a comparable IF input frequency range.

THE LOW-PHASE-NOISE LNB

All digital DTH modulation systems are based on the precise detection of phase and amplitude variations. The digital IRD must have the ability to detect these subtle variations at all times if the original picture and sound information is to be faithfully reconstructed. The ability of the IRD to detect these phase and amplitude variations can be hindered if the LO inside the LNB is unstable and produces excessive phase noise.

Phase noise is the result of short-term instability of the local oscillator or any other instantaneous frequency change of the local oscillator due to the quality factor of the oscillator components. An LNB employing low-

phase-noise technology contains a local oscillator that generates a frequency of higher spectral purity than is normally used in an LNB dedicated to the reception of analog TV signals.

Excessive LNB phase noise can cause the IRD to reproduce a picture that contains missing or erroneous components. The digital DTH industry refers to these distortions to the reconstructed picture as tiling, mosaics, and artifacts.

An LNB used to receive analog-based satellite TV signals may not be suitable for use in digital DTH receiving system. The installer should select an LNB employing low-phase-noise technology for all new installations as well as for any system replacements or upgrades.

LNB GAIN

Another important LNB specification is the amount of amplification or gain that each unit provides. This also is measured in decibels. The consumer-grade LNB commonly produces 50–65 dB of gain. The gain specification for any LNB is important for ensuring that the signal arriving at the IF input of the IRD is within a range of values in dBmV recommended by the IRD manufacturer. For example, an LNB with a gain of 65 dB is the logical choice for an installation where there is a long cable run between the outdoor and indoor units. The installer uses an LNB with 65 dB of gain to overcome signal attenuation through the cable, while an LNB with 50 dB of gain is preferred for installations with a short cable run between the outdoor and indoor units.

THE DUAL-POLARIZATION SINGLE-OUTPUT LNF

Many digital DTH systems can be ordered with an LNF that has a single IF output that places all transponders and polarizations from a given satellite onto a single coaxial cable. The so-called "stacked" LNF simultaneously receives both senses of orthogonal polarization and then downconverts one set of transponders to the 950–1,450 MHz IF band and the other set of transponders to the 1,550–2,050 MHz IF band. This eliminates the need for a feedhorn control cable and allows all of the available satellite channels to be fed to the indoor IRD by means of a single coaxial cable. Moreover, the dual-polarization single-output LNF also simplifies the installation of multiple set-top boxes in the home, with each IRD receiving all the available channels from a single coaxial cable connection.

KEY TECHNICAL TERMS

The following key technical terms were presented in this chapter. If you do not know the meaning of any term presented below, refer back to the place in this chapter where it was

presented or refer to the Glossary before performing the quick check exercises that follow.

Circular polarization

Feedhorn

Focal length

Heterodyne

Illumination taper

Isolator

Kelvin (K)

Linear polarization

Local oscillator

Noise figure

Noise temperature

Phase noise

Polarization

Thermal noise

QUICK CHECK EXERCISES

Check your comprehension of the contents of this chapter by answering the following questions and comparing your answers to the self-study examination key that appears in the Appendix.

Part I: True or False

_____ 1. The higher the noise temperature of the C-band LNB, the better its performance.

_____ 2. The gain of the LNB output is the ultimate figure of merit for calculating its performance.

_____ 3. The illumination taper of the feedhorn controls the signal contribution from various parts of the antenna's reflector.

____ 4. The focal length is the distance between the lip of the feedhorn and
the antenna's rim.

____ 5. Offset-fed antennas position the
feedhorn out of the path of the incoming signal.

Part II: Matching Questions

a. noise figure

b. *f/D* ratio

c. focal length

d. sidelobes

e. decibel

f. focal point

g. noise temperature

h. gain

6. The _____ is the measurement
from the _____ to the center of the
dish.

7. The _____ of an LNB is usually
50–60 dB.

8. Ku-band LNBs are rated according to
their _____ while C-band LNBs are
rated according to the _____.

9. The ratio of _____ to antenna
diameter is called the _____.

10. _____ of lower intensity are part of
every parabolic antenna's radiation
pattern.

Part III: Multiple Choice

11. You have just installed a C-band system
to receive a signal with an EIRP of 31

dBW. The correct combination of
antenna diameter and LNB noise
temperature is:

a. 1.8 m/60 K

b. 2.4 m/30 K

c. 3 m/90 K

d. 3 m/30 K

e. 2.4 m/90 K

12. You have just installed a Ku-band system
to receive a signal with an EIRP of 42
dBW. The correct combination of
antenna diameter and LNB noise
figure is:

a. 1.2 m/2.5 dB

b. 60 cm/1.2 dB

c. 60 cm/2.5 dB

d. 1.2 m/1.2 dB

e. 30 cm/1.2 dB

CALCULATING FOCAL LENGTH AND THE ANTENNA *F/D* RATIO

The focal length for any antenna can be computed if the diameter of the dish and its corresponding *f/D* ratio are known.

Focal length = Antenna diameter ×
f/D ratio

For example, the focal length of an antenna that is 3 m in diameter and has an *f/D*
ratio of 0.45 is:

$$3 \times 0.45 = 1.35 \text{ m or } 135 \text{ cm}$$

The *f/D* ratio can be calculated by dividing the focal length (*f*) by the antenna diameter (*D*). If the focal length is not known, the
focal length can be calculated as follows:

$$f = D2/16d$$

where $D2$ = the diameter of the paraboloid and d = the depth of the paraboloid from the reflector center to the rim of the dish.

CONVERTING NOISE TEMPERATURE (K) TO AN EQUIVALENT NOISE FIGURE (DB)

The following formula can be used to convert LNB noise temperature (K) to an equivalent noise figure (F) in decibels.

$F = 10 \log(K/290 + 1)$

For a noise temperature of 120 K:

$F = 10 \log (120/290 + 1)$
$F = 10 \log (1.41379)$
$F = 1.5038$ dB

INTERNET HYPERLINK REFERENCES

The Impact of Phase Noise on Satellite Television Reception.
http://www.calamp.com/support/14002.html

Isolated LNA and LNB Questions.
http://www.calamp.com/support/14003.html

Feedhorn Troubleshooting Guide.
http://www.chaparral.net/ts_feeds.htm

Chapter 5

Satellite Receiving Antennas

The satellite dish is a parabola of revolution, that is, a surface having the shape of a parabola rotated about its axis of symmetry (Figures 5–1 and 5–2). The resulting paraboloid shares one key property of optical lenses: it is able to form an image of whatever object is placed in front of it. The largest optical and radio telescopes employ the parabolic reflector to gather and concentrate electromagnetic radiation. Any antenna surface irregularities or any departure from the precise parabolic shape will degrade the image resolution. As is more often the case, however, low-resolution performance is the result of the installer's failure to grasp the importance of using good antenna assembly techniques.

The parabolic reflector receives externally generated noise along with the desired signal. When the satellite dish tilts up towards the "cold" sky, the antenna noise temperature is at its lowest level. If the antenna must tilt downward to receive a low-elevation satellite, however, the antenna's noise temperature will

increase dramatically because it is now able to intercept the "hot" noise temperature of the Earth (Figure 5–3). The actual amount of noise increase in this case is a function of antenna f/D ratio and diameter. Minimum antenna elevation angles of 5 degrees, for C-band, and 10 degrees, for Ku-band, above the site location's horizon usually are recommended.

DISH MATERIALS AND CONSTRUCTION

The reflector's surface material must be constructed out of metal in order to reflect the incoming microwave signals. Some antenna reflectors appear to be manufactured out of plastic or fiberglass; however, these dishes actually have an embedded metal mesh material that reflects the incoming satellite signals to the front and center of the dish.

The solid one-piece metal antenna is most always the dish with the best perfor-

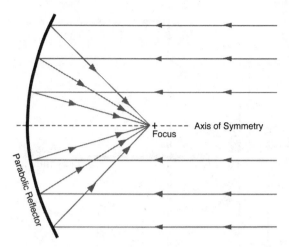

Figure 5–1 The parabolic curve has the property of reflecting all incident rays arriving along the antenna reflector's axis of symmetry to a common focus located to the front and center.

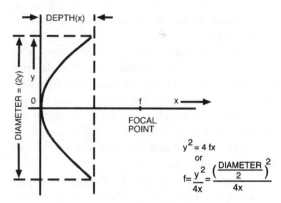

Figure 5–2 Calculation of the parabolic curve. (Courtesy of Taylor Howard.)

mance characteristics because there can be no assembly errors and the reflector normally will maintain its precise shape over the lifetime of the system. Solid petal antennas constructed

out of four or more metal panels are generally the next best performance value, as potential assembly errors are limited to variations along the seams between panels. The installer can visually inspect these seams during assembly to ensure that there are no variations in the surface curve from one petal to the next. Installation errors almost never occur when this type

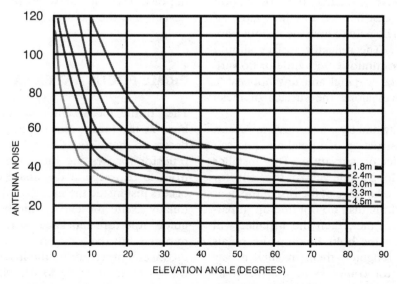

Figure 5–3 Antenna noise temperature is a function of the antenna f/D ratio and diameter as well as the elevation angle of the dish as it points toward the geostationary satellite's location in the sky.

of antenna is assembled face-down on a flat, level surface.

Both one-piece and petal antennas also are available in a perforated form. The desired diameter of the perforation holes is a function of signal wavelength: too small to pass or resonate with the wavelength of the incoming microwave signals, but large enough to pass light in order to minimize the antenna's environmental impact.

Mesh antennas are the most susceptible to construction errors. The two-part construction process consists of the building of a support frame, onto which is laid flexible mesh panels. The installer attaches the mesh material to the frame using a series of metal clips or sheet metal screws. This type of antenna is susceptible to construction errors because it must be assembled face up with no level surface to act as a jig to hold the reflector in place.

Mesh antennas also are highly susceptible to environmental effects. Heavy windstorms, for example, can loosen the clips holding the mesh to the frame and distort the curve from its original shape, or even blow out one or more of the mesh panels.

The installer should examine the antenna at intervals during the installation process. In the case of petal antennas, close attention should be paid to how the various panels fit together. The reflector surface should appear to be continuous, with little or no variation from petal to petal and few noticeable bumps or waves along the surface of mesh antennas.

Antenna symmetry is also very important. Improper construction of a petal antenna can warp the reflector curvature. The installer should sight along a side view of the reflector from the near to far edge of the rim. If the near and far rims of the dish do not line up in parallel with each other, then the installer will need to loosen the bolts holding the petals together and retighten them in such a way that the reflector conforms to the manufacturer's intended shape. Another way to detect

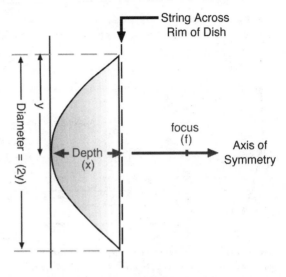

Figure 5–4 Stringing the dish is one method of visually verifying correct antenna construction. It also allows the installer to measure the antenna's depth.

a warped dish is to run a series of strings across the face of the antenna (Figure 5–4). All strings should lightly touch over the center of the dish. Any gaps between strings indicate a deviation in the parabolic curve.

PRIME FOCUS ANTENNAS

The "prime focus" antenna places the feedhorn at the focal point of the paraboloid reflector so that it looks back into the dish. The reflector focuses the incoming planar wavefront to converge at the phase center, which occurs just inside the mouth of the feed, to optimally excite the chosen mode in the waveguide. Aperture blockage by the feedhorn is minimal, but the location of the focus can be inconvenient, requiring the installer to mount the LNB also at prime focus, where it is inaccessible for adjustment and exposed to the direct and reflected rays of the sun as well as other adverse environmental effects.

Prime focus antennas are easy to construct and point toward the desired satellite. However, there are two main design disadvantages. The feedhorn and feed support structure block part of the reflector surface, and the feedhorn must look back at the dish at such an angle that it can also intercept noise from the "hot" Earth located directly behind the reflector. The way that the feedhorn illuminates the antenna must be tapered so that the noise contribution is minimized, as the feed looks outward toward the rim of the dish. This design necessity acts to reduce the maximum efficiency level that the antenna can attain. This is why prime focus antennas typically only achieve an efficiency level of 55–60 percent.

Prime focus antennas use two different types of feedhorn support bracket. A bracket with multiple support legs (either three or four legs) provides a rigid support structure for the feedhorn and LNB, placing them over the center of the dish and at the distance specified by the manufacturer. The main disadvantage of this structural approach is that it may be difficult to make minor variations in the focal length, that is, the distance from the reflector center to the lip of the feed opening.

The buttonhook structural design uses a single support member to position the LNB and feedhorn. This tubular leg usually can be slid in and out of a clamp or bracket at the center of the dish, allowing the installer to fine-tune the focal length quickly and easily. However, the buttonhook support may not always position the feed at the precise center of the dish, especially in those instances when the feedhorn is weighted down by more than one LNB.

Motorized dishes may experience feedhorn movement when the antenna moves from one satellite to the next; heavy windstorms can also temporarily move the feedhorn away from the antenna's focus. Guy-wire kits are available that the installer can use to provide additional structural rigidity to the buttonhook support if required for a given installation.

OFFSET-FED ANTENNAS

One oval dish design that is the antenna of choice for most digital DTH satellite TV service providers is called the offset-fed antenna (Figure 5–5). Here the manufacturer uses a smaller subsection of the same paraboloid used to produce prime focus antennas (see Figure 5–6), but with a major axis in the north/south direction, and a smaller minor axis in the east/west direction.

The offset paraboloid eliminates aperture blockage, reduces antenna noise temperature, and resists the accumulation of ice and snow by placing the feed below the reflector and angling it upwards. In this case, the reflector acts as if it were a portion of a much larger paraboloid. But because only a portion of this imaginary reflector exists, the feed is designed just to illuminate that portion. The offset-fed

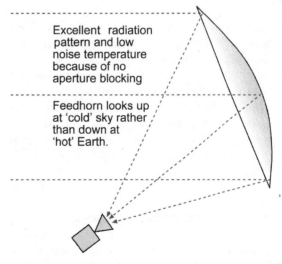

Excellent radiation pattern and low noise temperature because of no aperture blocking

Feedhorn looks up at 'cold' sky rather than down at 'hot' Earth.

Figure 5–5 *Schematic view of offset-fed antenna geometry.*

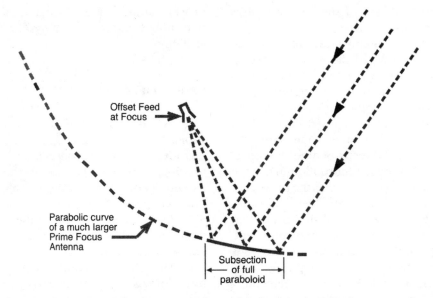

Figure 5–6 *The curve of the offset-fed antenna is actually a subsection of a paraboloid.*

antenna then performs just as it would as a part of the larger dish, and directs its beam exactly the same way.

The offset-fed antenna design offers several distinct advantages over its prime focus counterparts. There is no feedhorn blockage, an important consideration when the antenna aperture is less than one meter in diameter. Moreover, antenna designers can reconfigure the required antenna aperture as a flatter, more nearly vertical reflector, with the added advantage of pointing the feed skywards, away from the hot-noise source of the Earth. Because of these advantages, the offset-fed antenna can achieve higher efficiency levels than prime focus antennas normally attain, usually in the 70 percent range.

THE CASSEGRAIN ANTENNA

The cassegrain is a dual-reflector antenna design that primarily is employed at large-aperture uplink earth stations and cable TV head ends. The cassegrain improves aperture effi-ciency beyond the 55–60 percent that typically is achieved by the prime-focus antennas previously described by matching the feedhorn illumination profile more closely to the antenna aperture. It is characterized by a convex subreflector and a larger feed aperture than the prime focus antenna requires.

As with the prime focus dish, the cassegrain antenna's view of the satellite is partially obstructed, in this case by the subreflector (Figure 5–7). The subreflector diameter must be kept small to minimize blockage, but larger than about five wavelengths to minimize diffraction effects. Because of this five-wavelength limitation, the cassegrain design approach is not employed for C-band antennas that are smaller than 5 meters in diameter.

When the diameter of the main reflector exceeds 5 meters, however, the amount of subreflector blockage represents only a small percentage of the main reflector's total capture area.

This innovative dual reflector design allows the designer to approximate more closely the ideal of even illumination across the pro-

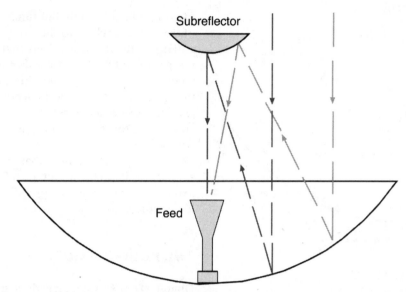

Figure 5–7 *Cassegrain antenna geometry. The cassegrain antenna's view of the satellite is partially obstructed by the subreflector. The subreflector diameter must be kept small to minimize blockage, but larger than about five wavelengths at the operating frequency to minimize diffraction effects.*

file of the main reflector without encountering appreciable noise beyond the antenna's rim. To accomplish this, the designer must alter the profile of the antenna's subreflector so it no longer is a true hyperboloid of revolution. This variance of shape alters the feed pattern on the main reflector to favor the region towards the rim, but with a rapid fall-off beyond the rim.

One side effect of this subtle change is that it also destroys the equality of all ray paths from the feedhorn to the far-field region so that waves reflected from different parts of the main reflector no longer arrive at the feedhorn in phase. To compensate for this, the designer must use a computer-derived profile to alter the main reflector from a parabola of revolution. This slight alteration is just enough to bring all points of the aperture back into phase inside the opening of the feedhorn. The result is an increased aperture efficiency of up to 78 percent, or a 1.5-dB increase in gain for a given antenna diameter. The precise manufacturing tolerances required, however, increase the cost of production and add to the complexity of the installation process.

THE SPHERICAL ANTENNA

Spherical antennas primarily are used for commercial SMATV and cable installations where the customer wishes to simultaneously receive multiple satellites with a single antenna. The spherical antenna design creates multiple focal points located to the front and center of the reflector, one for each available satellite. The curvature of the reflector is such that if it were extended it outward far enough along both axes it would become a sphere (Figure 5–8).

The planar wavefront emanating from a communications satellite can be described as a series of parallel rays. As we already have seen, if these rays approach in parallel to the paraboloid's axis of symmetry, they are reflected to a well-defined prime focus (F). This

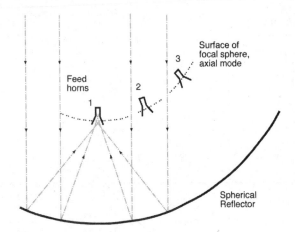

Figure 5–8 Spherical antenna geometry.

is the property of the paraboloid that makes it an excellent design choice for a microwave antenna.

The spherical dish, however, can be best visualized as a part of an imaginary ball, with its center at C, and where C is twice as distant from the spherical reflector as F would be on a parabolic reflector. Moreover, the focusing can best be described as an imperfect zone. If this focal cloud is small enough in comparison to the wavelength of the signals being received, the spherical reflector remains a useable antenna, but one with slightly less gain in comparison to its prime focus counterpart.

If the plane wavefront moves about 20 degrees off axis, the spherical antenna's main beam also can be steered simply by moving the feedhorn or by adding a second feedhorn at the new focus. The reflector effectively develops a new axis of symmetry upon which this new focus is centered.

The spherical antenna is a section of a sphere that potentially has any number of axes, with no one axis having supremacy over any other. Each axis represents a radius of the sphere and therefore has exactly the same relationship to the reflector as any other axis.

The antenna designer can achieve a good

focusing effect by only illuminating a portion of the reflector close to the axis. The resulting antenna will produce a gain that approaches the gain produced by a paraboloid with a diameter that is equal to the illuminated area of the spherical. The closer the feedhorn is to the axis, the more accurately the rays will be focused and the greater the antenna efficiency will be.

Most spherical antennas will work reasonably well up to about [±]20 degrees off axis. After that, defocusing will occur, which lowers the gain performance of the reflector.

THE PLANAR ARRAY

Digital DTH service providers in Japan and elsewhere have elected to use an alternative antenna design called the planar array, a flat antenna that does not rely on the reflective principles used by all antennas previously described. Instead, a gridwork of tiny elements is embedded into the antenna's surface. These elements have a size and shape that causes them to resonate with the incoming microwave signals. A spiderweb of feed lines is used to interconnect all the resonant elements in such a way that their signal contributions are all combined in phase at a single terminal located at the center of the array that connects directly to the LNB.

Planar arrays are relatively unobtrusive: there is no feedhorn and the LNB is located to the rear of the antenna out of sight. Since these antennas are most always dedicated to the reception of a single satellite or constellation of colocated satellites, they can be mounted in a fixed position on an outer wall or rooftop.

One main disadvantage of the planar array is its limited frequency bandwidth, 500 MHz, in comparison to parabolic antennas. The paraboloid is a broadband device that in many instances can be used to receive S, C, and Ku-band satellite signals. Another disadvantage of the planar array is the high manufacturing costs involved. The retail price of a planar array is more than four times the cost of a feedhorn and parabolic reflector with equivalent signal amplification characteristics.

ANTENNA GAIN AND *G/T*

The gain of a satellite antenna is the measurement of its ability to amplify the incoming microwave signals. Gain—which is expressed in decibels, or dB—is primarily a function of antenna capture area or aperture: the larger the antenna aperture, the higher the antenna gain. Gain also is directly related to the antenna's beam width, the narrow corridor along which the antenna looks up at the sky. (See Figure 5–9.)

The antenna's efficiency rating is the percentage of signal captured by the parabolic reflector that the feedhorn actually receives. As previously outlined, the feedhorn's illumination of the outer portion of the dish is attenuated or tapered, which leads one to conclude that antenna gain is not as important a factor as it might first appear to be.

The ultimate figure of merit for all receiving antennas is the *G/T* (pronounced "*G* over *T*"); that is, the gain of the antenna (in decibels) minus the noise temperature of the receiving system (also in decibels). A typical C-band DTH system has a *G/T* of about 20 dB/K, while most Ku-band DTH systems have a *G/T* of about 12.7 dB/K. The higher the satellite power, the lower the *G/T* value that will be needed at the receiving system down on the ground.

The noise value (*T*) primarily comes from two sources. The antenna noise is a function of the amount of noise that the feedhorn sees as it looks over the rim toward the hot Earth (which has a noise temperature of 290 K). Antenna noise generally ranges between 30 and 50 K.

The noise contribution of the circuits inside the LNB is the other major source of concern. C-band LNB performance now ranges as low as 20 K. If we add a typical antenna/feed noise of 40 K to LNB noise of 35 K, we get a system noise temperature of 75 K. Ten times the logarithm of 75 K equals a *T* of 18.8 dB. A typical 1.8-m diameter C-band antenna will

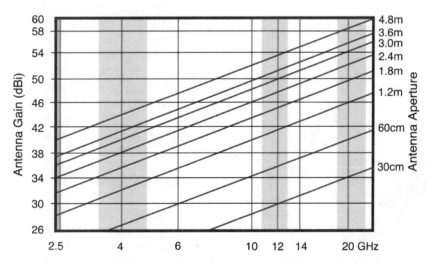

Figure 5–9 *Antenna gain (G) is a function of satellite frequency, antenna diameter and aperture efficiency. G = 10 log (4πAg/λ²), where* A *= the effective antenna area, which is* πR² *for a circular paraboloid;* g *= the aperture efficiency expressed as a decimal;* λ *= the signal wavelength; and* π *= 3.14159.*

produce a gain of 38 dB. Therefore, the G/T of the system just outlined would be (G) 38 dB minus (T) 18.8 equals 19.2 dB/K.

ANTENNA F/D RATIO

The f/D ratio of the antenna is the ratio of focal length to dish diameter, measured in the same units (Figure 5–10). A paraboloid reflector that is 3 m in diameter and with a focal length of 1.26 m therefore will exhibit an f/D of 0.42. The f/D ratio selected by the antenna designer also determines the depth of the dish itself, that is, the amount of contour or "wrap-around" of the paraboloid within its fixed diameter. A long-focus (high f/D) paraboloid reflector will have a shallow contour, while a short-focus paraboloid reflector resembles a deep bowl. The deepest reflectors have a f/D ratio of 0.25. This places the focal point directly in the plane of the antenna aperture.

An antenna design with a large value of f/D requires a feedhorn that has a narrower beam width, so that the edge illumination of the antenna can be maintained. This typically is between 10 and 15 dB below the value pro-

duced at the center of the reflector. Conversely, a small value of f/D will require a feedhorn with a wider beam width.

Paraboloid antennas that are 3 m or less in diameter (at 4 GHz) commonly use a 12-dB feed illumination taper, while larger antennas will use a 15-dB taper. The antenna designer must make a trade-off between antenna gain and noise temperature, balancing the entry of random noise due to feedhorn overillumination or low antenna elevation angle with the noise contribution of the antenna sidelobes in the antenna radiation pattern.

Although the long focal length employed by the shallow dish design increases the illumination of the reflector surface, there are distinct disadvantages to this design approach. Moreover, antenna noise increases as antenna elevation increases. Shallow dishes are more susceptible to intercepting Earth noise when pointing at low elevation angles. Finally, the shallow dish is more susceptible to picking up terrestrial interference from terrestrial microwave stations.

The deep-dish design trades off gain in order to lower antenna noise performance. The deep-dish design is an attractive alternative for locations that potentially may experience terrestrial interference (TI) problems or at installations that require low antenna elevation angles. The deep-dish design positions the feedhorn relatively close to the rim of the reflector. Therefore, the deep dish has a greater ability to shield the feedhorn from potential TI sources. However, the feedhorn is so close to the reflector that it cannot effectively illuminate the entire surface.

ANTENNA SIDELOBE REJECTION

The explosive growth in satellite telecommunications around the world is leading to closer spacing between satellites in geostationary orbit. Moreover, the very latest satellites are transmitting higher-powered signals. Both of these developments act to increase the poten-

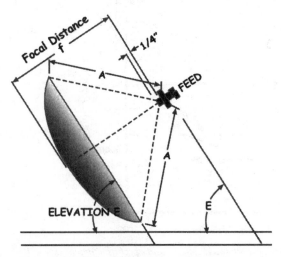

Figure 5–10 The feed must be precisely centered over the antenna reflector at the specified distance.

tial for interference between satellites in close proximity to one another.

The perfect parabolic antenna would only receive signals from the satellite at which it was pointed while rejecting all signals coming from other directions. In the real world, however, each antenna design will produce a main beam along the axis of symmetry as well as other lower-intensity beams that look out at adjacent angles. These beams of lower intensity are called sidelobes (Figure 5–11).

Antenna performance usually is measured at an antenna testing range, where a distant radiating point is rotated across the antenna's field of view while a varying pattern representing the antenna's gain in decibels is plotted on a graph (Figure 5–12). The antenna beam pattern that is produced in this manner shows the off-axis location for each sidelobe as well as the amount in decibels that each sidelobe is down from the antenna's (–3 dB points) main beam.

The goal of all satellite TV antenna manufacturers is to reduce the gain of these sidelobes to levels that are –15 to –18 dB or more below the gain of the main beam. This amount of sidelobe attenuation usually is enough to keep adjacent satellites from causing interference to reception of the desired satellite. The location off axis of each sidelobe also is a function of antenna diameter and signal frequency. The system designer either can select an antenna that is large enough to put the adjacent satellites in the first "null" of the antenna receiving pattern, or use an antenna with a first sidelobe that is at least –15 dB down from the main beam.

ANTENNA NOISE TEMPERATURE

All satellite antennas receive thermal noise along with the desired satellite signal. When the satellite dish tilts up towards the "cold" sky, its noise temperature is relatively low. If the antenna must tilt downward to receive a low elevation satellite, the antenna's noise temperature will increase in a dramatic way because the antenna sidelobes are now able to see the "hot" noise temperature of the Earth.

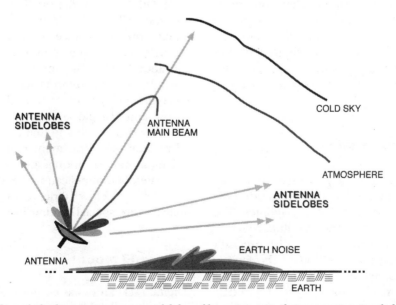

Figure 5–11 All paraboloid antennas generate sidelobes of lower intensity that can receive signals from sources other than the one at which the antenna is directed.

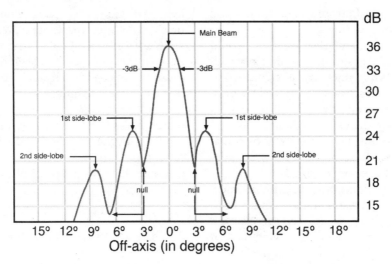

Figure 5–12 Antenna test pattern showing the sidelobe performance of a 60-cm dish.

The actual amount of noise increase in this case is a function of antenna diameter. Minimum antenna elevation angles of 5 degrees for C-band and 10 degrees for Ku-band usually are recommended.

Whenever low antenna elevation angles are required to receive the selected satellite or satellites, a deep dish usually will exhibit less pronounced sidelobes than a shallow dish of equivalent diameter and therefore will intercept less thermal noise at low antenna elevation angles.

THE ANTENNA MOUNT

All satellite antenna mounts incorporate adjustments that permit the installer to point the dish at the desired satellite or satellites. The mount that supports the satellite reflector must be capable of pointing the dish precisely and maintaining that position in the face of various environmental forces. For example, a 1.25-cm movement at the rim of a 1.5-m dish or a 2.5-cm movement at the rim of a 3-m dish

will cause a full degree of reflector movement. Therefore, the installer should be aware that small movements of the parabolic reflector could make the difference between excellent digital TV reception and no reception at all.

On occasion, the satellite or satellites of interest will be located near to the site location's horizon. To receive these satellites, the receiving antenna must tilt downward to a low elevation angle. The installer should always consult with the antenna manufacturer prior to commencing work at a site that will require a low antenna elevation angle to determine whether or not the antenna mount is capable of making this adjustment (sometimes called the minimum elevation angle). Some commercial antenna mounts cannot be adjusted to angles that are below 10 degrees of elevation, or may require special modifications to do so.

The Az/El Mount

Digital DTH antennas commonly have what is known as a fixed mount that is adjusted once at the time of installation and then left alone thereafter. This azimuth over elevation (Az/El) mount must be capable of making two independent adjustments to align the reflector in the direction of the selected satellite. The angular movement east, or west, in the site location's horizontal plane is called the azimuth; the angular movement or "tilt" up from the site location's horizon is called the elevation. For multiple satellite viewing, automation of the Az/El mount requires two motors to move the dish to the set of azimuth/elevation coordinates that corresponds to the location of the selected satellite, or colocated satellites, in the sky.

The horizontal/vertical orientation of the antenna and feed is relative to the site location's horizontal plane. The horizontal orientation of each satellite, however, is relative to the Earth's equatorial plane—the flat plane passing through the Earth at its equator and extending out into space—and to the vertical orientation of each satellite relative to the Earth's rotational pole.

From the perspective of the site location, the orientation of the incoming satellite signal rotates, or "skews," relative to the local horizontal plane for satellites that are located to the east or west of the site location's meridian. Any motorized antenna with an Az/El mount will therefore require a feedhorn that can make the necessary skew corrections to optimize the polarization setting of the receiving system.

The Polar Mount

The polar mount tracks the geostationary satellite arc by means of a single adjustment of rotation about the mount axis. This is called the actuation. The main advantage of the polar mount is that only one motor is required to actuate the dish.

Astronomers typically mount their optical and radio telescopes on what is called a true polar mount, which has an axis that is parallel to the Earth's pole of rotation. The orientation of the "true polar mount" axis works out with geometric precision to be in line with true north/south and inclined to the exact angular magnitude of the local latitude.

Although the true polar mount is precisely what astronomers need to track distant celestial objects such as planets, stars, and galactic clusters, it is not suitable for tracking relatively close objects such as communications satellites. The geostationary satellites, after all, are in close proximity to the Earth when compared to the vast distances between the Earth and other celestial bodies. For this reason, the polar mount axis must be modified so that it is inclined slightly toward the Earth's equator. This modification to the design of a true polar mount is called the "declination." Once set to the correct declination angle, the modified polar mount will pivot or "actuate" around its axis to sweep through the geostationary arc to acquire each satellite with an accuracy of within a few thousandths of a degree (see Figure 5–13).

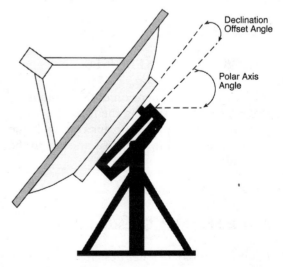

Figure 5–13 The modified polar mount.

KEY TECHNICAL TERMS

The following key technical terms were presented in this chapter. If you do not know the meaning of any term presented below, refer back to the place in this chapter where it was presented or refer to the Glossary before performing the quick check exercises that follow.

Antenna gain

Az/El mount

Cassegrain

Declination

f/D

Focal length

Focal point

G/T

Modified polar mount

Offset-fed antenna

Parabola

Planar array

Prime focus

Sidelobe

QUICK CHECK EXERCISES

Check your comprehension of the contents of this chapter by answering the following questions and comparing your answers with the self-study examination key that appears in the Appendix.

Part I: Matching Questions

Put the appropriate letter designation—a, b, c, d, etc.—for each term in the appropriate blank space in the matching sentence.

a. focal length

b. *C/N*

c. gain

d. *G/T*

e. focal point

f. parabolic

g. *f/D* ratio

h. beam width

i. declination

j. sidelobe

k. axis of symmetry

l. hour angle

1. A correction factor called the _____ must be incorporated into the modified mount to tilt the antenna downward toward the geostationary arc.

2. The _____ of an antenna is an expression of the signal amplification provided by the reflector.

3. The _____ is the measurement from the _____ to the center of the antenna reflector.

4. The figure of merit for any satellite receiving system is its _____ rating expressed in dB.

5. A _____ antenna has the ability to reflect all incident rays arriving along the antenna's _____ to a common location called the _____ that is located at the front and center of the dish.

6. The _____ of an antenna determines the area of the sky that is received by the antenna's main beam, while the _____ are secondary beams of lower intensity which are capable of picking up signals from satellites adjacent to the selected one.

Part II: True or False

____ 7. The gain of a parabolic antenna is the most important figure of merit for evaluating its performance.

____ 8. Antenna noise temperature is a function of the elevation angle of the dish.

____ 9. The primary advantage of a cassegrain antenna is that its subreflector blocks a lower percentage of the antenna's total surface area than the feedhorn blocks on a prime focus antenna.

____ 10. The G/T for any satellite TV receiving system is determined by dividing the antenna gain in decibels by the noise temperatures of the LNB and antenna.

CALCULATING ANTENNA GAIN

Expressed in decibels, the gain (G) of a paraboloid is a function of antenna diameter at a given frequency plus the efficiency of the antenna system, including the feedhorn.

$$G = 10 \log(4\pi Ag/\lambda 2),$$

where A = the effective antenna area, which is πR^2 for a circular paraboloid; g = the aperture efficiency expressed as a decimal; λ = the signal wavelength; and π = 3.14159.

G at 4,200 MHz of a 3-m dish when $g = 0.55$ is:

$G = 10 \log (0.55 \times (12.57 \times A))/(.0714)2$
$A = \pi R^2 = (3.14159 \times (1.5)2 = 7.07$
$G = 10 \log (0.55 \times 88.87)/(0.0714)2$
$G = 10 \log (48.88/0.005)$
$G = 10 \log 9775.7 = 39.9$ dB

Chapter 6

The Integrated
Receiver/Decoder (IRD)

Today's communication satellites relay a more diverse range of TV programming than ever before. The selection from among the various satellite receivers now on the market will therefore depend largely on the nature of the satellite TV program services that customers want to view from the comfort of their easy chairs. Some satellite TV programs are broadcast in the clear and are therefore considered "free to air" (FTA); others may be encrypted and only made available to subscribers living within a particular region or individual country.

There also is an important distinction to be made between satellite TV services that are transmitted in an analog format and those that are digitally compressed. Analog communication signals are electromagnetic waves of energy that vary in intensity (called amplitude modulation, or AM) and/or frequency (called frequency modulation, or FM). Digitally compressed TV signals, however, are broadcast in an alternate format consisting of a series of bi-nary digits or bits that correspond to the on (1) and off (0) states of computer logic circuitry. Only one manufacturer currently offers a satellite TV receiver that can process both analog and digital satellite TV signals. This is the 4DTV IRD that General Instrument has designed for the reception of analog and digital satellite TV services that are transmitted in the clear as well as those encrypted services using General Instrument's VideoCipher RS (analog) and DigiCipher (digital) encryption systems.

To fully understand how new digital products are changing the way that consumers view satellite TV, the following chapter compares and contrasts some of the features, options, and performance variables of analog satellite TV receivers and integrated receiver/decoders to digital IRDs (Figures 6–1 and 6–2). Toward the end of the chapter, information will be presented concerning how to troubleshoot digital DTH systems in the event that something ever goes wrong.

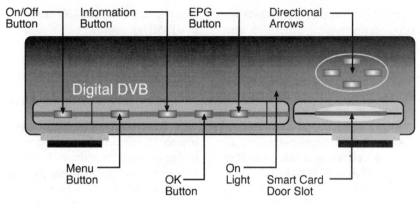

Figure 6–1 *A digital integrated receiver/decoder (IRD).*

ANALOG SATELLITE TV RECEIVERS

The cost of an analog satellite TV receiver depends on the number of features that it offers to the consumer. Some low-cost satellite receivers can display TV images that equal or even exceed the picture quality offered by more expensive units. The downside is that the operator will have to remember and execute the numerous little adjustments needed to tune into each and every satellite TV broadcast.

In some cases, the receiver may be a stand-alone unit that, by itself, can only receive programs from a single satellite. Most analog satellite TV receivers, however, also feature built-in antenna controllers that can steer a motorized dish from one satellite to any other.

The receiver's wireless remote control can be regarded as the keyboard, whilst the TV set serves as a computer screen which can display a variety of menus that the installer uses during the initial installation process. The TV viewer also can use the remote control to customize the unit to suit his or her specific viewing needs. Virtually all high-end satellite TV receivers are actually task-specific computers that contain advanced microprocessors and memory storage circuits. Each receiver comes with factory-installed software programs that

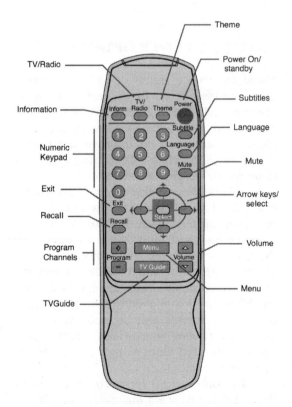

Figure 6–2 *Infrared remote control.*

automatically perform all the various tasks required to receive the available satellites and satellite TV services.

All satellite TV receivers are designed to maximize the strength of the incoming satellite signal or carrier while at the same time limiting the amount of noise either contributed by external sources or generated by the satellite TV system's internal electronic circuitry.

The figure of merit for an analog satellite TV receiver is defined as a threshold point expressed in decibels (dB) at a specific carrier to noise ratio (*C/N*) (Figure 6–3). As the receiving system's *C/N* approaches this threshold point, comet-tailed dots of impulse noise, called "sparklies," begin to appear in the TV picture. The lower the receiver's threshold rating, the better it will operate under low signal receiving conditions.

The mathematical relationship between *C/N* and *S/N* can be computed if we first con-

vert the *C/N* (dB) to *C/No* (noise power density in dB-MHz).

$$C/No = C/N - 10\log /Bn$$

where Bn = receiver IF bandwidth.

Video S/N = C/No + 22.6 (dB)

Receiver threshold ratings ranging from 6.5 to 10 dB *C/N* commonly are encountered in the product literature. The problem with relying on this specification as a benchmark for receiver evaluation is that not all manufacturers measure the threshold performance of their products in the same way. The best method for evaluating satellite receiver performance is to connect the unit to a dish of the same diameter as the one that will be installed at the customer's home and actually see how it performs while receiving several different satellite services.

ANALOG RECEIVER OPTIONS

There are a wide variety of optional features that may be offered by different analog receiver models. Although none of these options is essential to satellite TV viewing, they may allow the customer to improve the performance of the satellite TV receiving system or dramatically simplify its operation.

Filters. Many analog receivers offer noise reduction filters that can be used to improve reception of the weaker satellite TV channels (Figures 6–4 and 6–5). Keep in mind that these filters lower the unit's threshold point at the expense of video fidelity. When the noise reduction filter is engaged, for example, sparklies may be removed from the image, but other impairments will be noticeable, such as picture tearing when video graphics containing saturated blues or reds appear on the TV screen.

In some areas, terrestrial communications services transmit microwave signals that can adversely affect satellite TV reception. This phenomenon is known as terrestrial interfer-

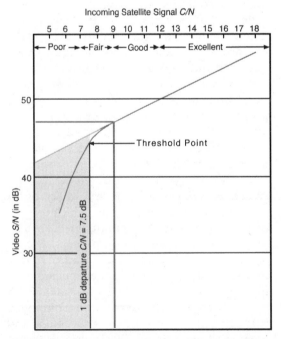

Figure 6–3 *Analog satellite TV receiver threshold is the point at which the relationship between* C/N *and* S/N *becomes nonlinear.*

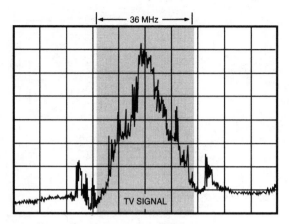

Figure 6–4 Spectral display of unfiltered reception of an analog satellite TV signal.

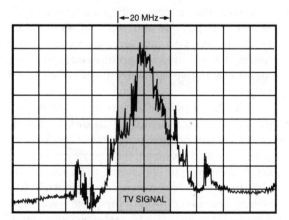

Figure 6–5 Spectral display with bandwidth filter engaged.

ence (TI). Some receivers provide special TI filters that can be used to reduce or eliminate TI problems that may plague reception of certain satellite TV services.

Automatic VCR timer. This is a desirable feature for anyone who would like to automatically tape satellite TV broadcasts for later viewing. In this case, the receiver must have an internal clock for keeping track of time. The receiver can then be programmed to receive a specific satellite and TV channel at the selected

time, regardless of whether the viewer is home or not. Most units offering this feature can be programmed to receive multiple events over a period ranging from two to four weeks. The VCR's clock must be synchronized with the satellite receiver's clock, and the VCR must be programmed to record at the same times that have been programmed into the satellite receiver's VCR timer.

Auto peak. The auto peak feature allows the operator to fine-tune system performance if there has been a noticeable degradation in the quality of reception from a given satellite or satellite TV service. For example, a windstorm or wayward soccer ball may have pushed the dish slightly away from its normal alignment. When the auto peak feature is engaged, the receiver will measure the strength of the incoming signal while it moves the dish back and forth and fine-tunes the channel frequency and polarization settings. The receiver will automatically select the settings at which it obtained peak signal reception. The operator can accept the new settings and store them in the receiver's memory or return to the settings previously stored in memory. This auto peak feature is not practical for systems that use antennas less than 3 m in diameter. On smaller dish systems, the receiver may lock onto a signal coming from an adjacent satellite and viewers will be lost in space before they know it.

Parental lockout. This feature allows the operator to designate certain satellite TV channels as off-limits to children. An on-screen menu allows the receiver operator to program the receiver to pick up these channels only after the viewer has entered a secret password. On some IRDs, the parental lockout feature can also be used to prevent inquisitive children from accidentally changing the primary tuning parameters of the system.

Favorite channels. Some of the high-end receivers allow the operator to create a customized list of favorite video and audio services that can be used for program selection. In this case, the TV viewer does not have to remember the name of any satellite or the channel number of a given satellite TV service. The operator merely displays the favorite channel list on the TV screen, then selects the desired service, and the satellite receiver does the rest.

ANALOG SATELLITE TV IRDS

All of the features and options just mentioned for analog TV receivers are also provided by any top-of-the-line IRD. The main difference between an IRD and a receiver is that the integrated receiver/decoder also contains a module that can decode those TV program services that have been encrypted to prevent unauthorized reception. For example, a movie service may employ the VideoCipher encryption system, while other analog TV programmers in the region use B-MAC, Nagravision, Smart-Crypt, Syster, VideoCrypt, or some other encryption system. Since none of these encryption systems are mutually compatible, the installer must take care to select an IRD that is fully compatible with the encrypted program service or services that the customer wants to view. Customers also must be aware that they will have to pay a subscription fee before the programmer will allow them to receive an encrypted satellite TV channel or a package of services.

WHAT IS ENCRYPTION?

Encryption is an electronic method of securing the video and audio of any TV program so that satellite, cable, and broadcast TV services can maintain control over the distribution of their signals. To receive encrypted or "scrambled" TV services, cable and SMATV system operators, hotel chains, private satellite networks, and home dish owners must possess a compatible decoder that can sense the presence of the encrypted TV signal and then automatically decode the pictures and sound (Figure 6–6).

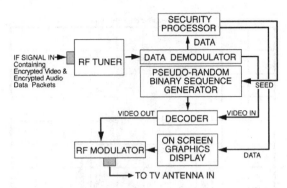

Figure 6–6 Decoder encryption components.

Premium program services purchase the rights to movies from film production companies with the understanding that every individual will pay for the right to view them. Programmers also are very concerned about hotels, bars, and other commercial establishments that derive monetary benefit from signal piracy.

Within a particular region, program producers may license more than one broadcast outlet for use of their programs. The program producer may require that broadcasters encrypt their signals whenever the broadcaster airs the producer's copyrighted material. This strictly limits reception of the programming to the market for which each broadcaster is licensed. In some areas of the world, satellite broadcasters periodically must switch from a free-to-air to an encrypted transmission mode whenever required under their respective agreements with the program copyright owners.

Each IRD contains a unique numerical address number that is installed at the factory. The satellite TV programmer's authorization center sends a coded conditional access message over the satellite that includes this unique address. This authorization message can turn on an individual IRD so that it can receive a particular service or group of services, or turn off an IRD in the event that the subscriber fails to pay the required monthly subscription fee.

Moreover, the authorization center can use this addressable feature to selectively turn off and on large groups of decoders. Group IRD control is used to selectively "black out" TV events, such as a live championship boxing match, in certain countries for which the programmer does not own the distribution rights.

Some encryption systems also require the use of a plastic security card, called a smart card, which plugs into a conditional access module that has a card reader slot built into the front panel of the IRD (Figure 6–7). The security card consolidates multiple encryption features into a single, electronic chip that contains a special set of mathematical algorithms—electronic keys that must reside in a decoder's circuitry before signal decoding can take place. Once it is plugged into the IRD, the conditional access module reads the data off of the smart card.

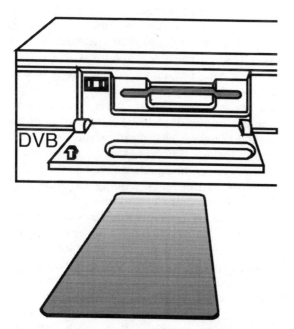

Figure 6–7 IRD conditional access systems feature a smart card reader, also referred to as a conditional access (CA) module.

ANALOG ENCRYPTION

Early video encryption systems relied on a couple of simple encoding techniques to secure the video from unauthorized reception. These so-called "soft" encryption systems removed the standard vertical and horizontal sync pulses from each video frame and/or inverted the video waveform to create a negative image. Some satellite TV viewers, however, quickly figured out how to generate the missing sync pulses and "right" the inverted video.

"Hard" video encryption is accomplished through the use of three different techniques: line translation, line delay, and line shuffling. With line translation, segments of each digitized line of video are sampled by the encoder and converted into digital values. The digitized line segments are then cut and rotated so that the segments within each line are shuffled out of order and reassembled at either side of the cut points. Each line has different cut points; all vertical information in the picture is broken up—stepped back and forth across the screen with each line—and in a sequence that changes from field to field. The decoder performs the complementary cut-and-rotate operation, patching each line back together at the correct point to reconstruct the original picture.

With line delay, the start points for each video line are delayed by varying intervals from their usual start time. Line shuffling rearranges the order of the lines presented on the TV screen in a pseudorandom fashion, much as a dealer randomly shuffles a deck of cards prior to dealing.

BASIC ENCRYPTION ELEMENTS

Analog encryption systems—irrespective of whether they employ active line translation, line shuffling, or line delay encryption methods—need a means for selecting at random the implementation points governing the presentation of each line of video. The up encoder contains a device called a pseudo-random binary sequence (PRBS) generator that produces the sequence of numbers used to control the selection of these implementation points. The result is a scrambled video display having little or no viewing value whatsoever. The IRD receives the encrypted lines of video, reassembles them into the correct sequence, and then outputs an error-free picture to the TV set.

The IRD must have instantaneous access to the location of these points used to alter each line of video from its original unencrypted state. The IRD obtains this information from its own internal PBRS generator, one that has characteristics identical to the encoder's PRBS.

Another essential component of any encryption system is a randomly generated control algorithm: a numerical "key" that the encoder can frequently change at will. Because the encoder interrupts and restarts the final control algorithm periodically, the decoder must receive precise synchronization instructions from the encoder. This set of synchronization instructions, which dynamically locks the twin PRBS generators, is called the "seed." The encoder sends the seed, which also is changed periodically, to all subscribing IRDs over the satellite link along with the program service or services.

When used with analog TV encryption systems, the seed is contained in the video signal's vertical blanking interval (VBI). The seed also is encrypted to ensure that only authorized subscribers will gain access to it. The security card contains the electronic keys that will unlock the seed. The smart card stores the appropriate algorithm inside a tamper-proof solid-state electronic chip that is embedded within it.

DIGITAL IRDS

All digital DTH transmissions are more or less compliant with a new MPEG-based digital video broadcasting (DVB) standard that initially was developed in Europe and has since been adopted by numerous satellite TV broadcasters around the world. Unlike analog satellite TV channels, which exist as stand-alone services on dedicated transponders, digital satellite TV services most often are part of a program package consisting of a group of channels. Through digital compression, multiple TV and audio signals are combined or "multiplexed" into a unified digital bitstream that shares common encryption, electronic program guide (EPG), and service information components. This multiplex is then transmitted through one or more satellite transponders on an individual satellite. The sharing of common elements between various satellite TV services is what creates a unified digital "bouquet".

Some digital bouquets are multichannel free-to-air services, while others are encrypted to prevent unauthorized reception. To receive the latter, a compatible digital IRD (Figure 6–8), subscription authorization, and, in most cases, a smart card are required before the customer can receive the available programming.

The MPEG-2 and DVB standards do *not* specify any particular conditional access (CA) system; each programmer therefore is free to choose the CA system that best fits its particular needs. There also are other restrictions to the use of one digital IRD to receive multiple digital bouquets sharing the same CA system. These include differences in the transmission rates between bouquets, as well as the subtle changes to the MPEG coding structure that each programmer elects to make.

Prior to transmission via satellite, the DVB digital bitstream is first converted from binary digits to symbols, with two bits combining to make each symbol. The symbol rate for a DVB-compliant satellite transmission may vary from just a few million symbols per second to a rate of 30 Msym/s, depending on the number of services that each satellite transponder contains. The digital IRD must be capable of tuning to the exact symbol rate

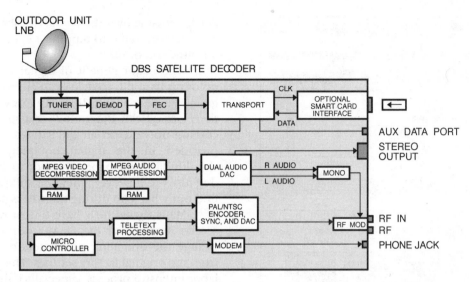

Figure 6–8 *Block diagram of a digital integrated receiver/decoder.*

used for a particular DVB transmission before it can decode the signal. For example, digital IRDs with a tuning range of 15 to 30 Msym/s cannot decode digital DTH signals that are transmitted at lower symbol rates.

Furthermore, every digital DTH bouquet uses forward error correction (FEC) to improve the robustness of their signals. FEC rates of 1/2, 2/3, 3/4, 4/5, 5/6, and 7/8 (original information symbols/FEC corrected symbols) are used by various digital DTH systems around the world. The digital IRD must be capable of either tuning to or automatically selecting the FEC rate used by the service that it is attempting to receive (Figure 6–9).

There are major differences between the features and options offered by analog IRDs and their digital counterparts. Most digital IRDs are programmed at the factory to receive a digital DTH bouquet from just one satellite. This initial setup includes the satellite transponder's center frequency and polarization format, as well as the service provider's symbol and FEC rates. Therefore, the IRD has no means for switching from one satellite to another, nor is the remote control equipped

with buttons for changing satellite transponders or transponder polarization. The good news for TV viewers is that operating a digital IRD is much like operating a regular TV set: flip through the channels, kick back, and enjoy.

Once installed, the digital IRD will tune automatically to a factory-programmed "default transponder" and access the EPG, service information, and CA data that it needs before it can begin delivering signals to the TV set. The service information data transmitted over the satellite provides picture (PID) and sound (SID) identification numbers that assist the IRD in locating every satellite TV and audio service in the bouquet. The TV viewer merely has to select the desired video or audio service, and the digital IRD does the rest. If for any reason the location of any digital satellite service in the bouquet ever changes transponders, the service information data transmitted over the satellite will automatically notify the IRD. Any changes of this nature are totally transparent to the TV viewer.

Digital IRDs also do not have any built-in noise reduction or TI filters for the viewe

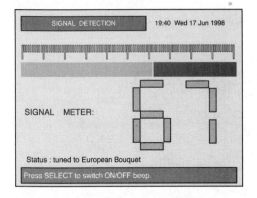

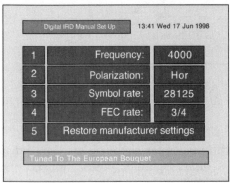

Figure 6–9 Transponder center frequency, polarization, symbol rate, and FEC rate for the digital bouquet must be set to the correct values before the IRD can receive the services. The correct LNB local oscillator (LO) frequency also must be programmed into the IRD, either at the factory or by the satellite system installer.

adjust. This is due to the major differences between analog and digital DTH transmissions. Narrowing the bandwidth of an analog TV transmission trades off some picture quality for an increase in system *C/N*. With a digital signal, however, we cannot narrow the bandwidth without losing essential components of the incoming signal.

The threshold of a digital IRD also is totally different from that of its analog counterpart (Figure 6–10). As the *C/N* of an analog TV signal falls below the IRD threshold point, the picture rapidly becomes noisy, but is still visible. With a digital signal, however, the IRD displays either a perfect TV picture, or no picture

at all. Crossing over the digital threshold point is therefore similar to flipping a light switch— it's either on or off.

One major benefit of digital DTH technology is its ability to deliver an electronic program guide to the IRD. TV viewers can therefore quickly determine what services are coming up next on all the satellite channels provided in the digital bouquet. Viewers shouldn't toss the paper TV guide to the family dog just yet, however. They will still need it to find out what's on during the weeks ahead.

Many digital IRDs have one or more on-screen menus that the viewer can access to change the operating parameters of the IRD from the original factory settings. This can be a labor-intensive process, especially if the operator needs to switch satellite frequency bands. Although playing with the settings of a digital IRD is now in vogue among satellite hobbyists around the globe, few satellite TV viewers are likely to attempt this very often, if at all. Future products will be soon coming our way that will be more versatile and user friendly. Meanwhile, be sure that the digital IRD is fully compatible with the digital bouquet that offers the programming to which the customer wishes to subscribe.

ENCRYPTING THE MPEG'S CUBE

The MPEG-2 digital transport layer consists of addressable packets that can be subjected to algorithms in a greater number of ways than what is possible when working in an analog domain. For example, the conditional access (CA) system for a digital DTH transmission does not have the bandwidth constraints imposed on analog encryption systems, which must address each IRD by means of data that is inserted into the limited confines of the analog TV signal's horizontal or vertical blanking intervals.

With analog encryption (Figure 6–11), the viewer will usually encounter a jumble of squiggles or geometric patterns running

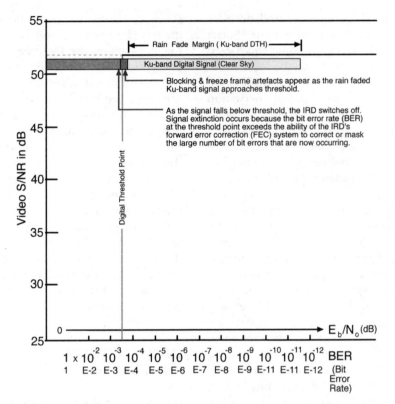

Figure 6–10 The threshold for a digital satellite receiver is defined as occurring at a particular bit error rate (BER). Reduction in the incoming signal level (E_b/N_o, expressed in decibels) has no effect on video S/N. Instead, video signal quality is determined by the bit rate and frame resolution assigned to each individual TV service within the MPEG-2 digital multiplex. Blocking and freeze frame artifacts, however, will occur as a rain-faded Ku-band satellite signal approaches the digital threshold, just before the digital IRD switches off.

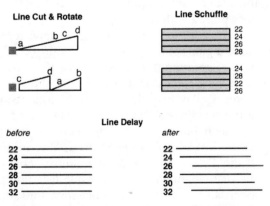

Figure 6–11 Analog satellite TV encryption techniques.

through the TV screen that identifies the presence of an encrypted TV service. MPEG-2 signals, however, appear to the uninitiated IRD as virtually indistinguishable from random noise. A spectrum analyzer or other signal measurement device can be used to detect the presence of a signal. However, there is no way to visually determine from a spectral display or a signal-level readout whether or not any digital signal contains video information.

Today's MPEG-2 conditional access systems share many of the key aspects used by their analog counterparts. For example, pseudorandom binary sequence generators are used to generate electronic keys. The precise synchronization of the encoder and decoder is another an essential requirement. Smart cards, as well as the corresponding smart card readers (also called conditional access modules), which are built into all digital IRDs, are also integral components of digital encryption systems.

In fact, some digital DTH services use a specialized version of a conditional access system that is already in use elsewhere to encrypt analog satellite TV broadcasts. For example, DirecTV uses a version of the VideoCrypt conditional access system developed by News Datacom for analog TV use in Europe. The VideoCipher RS (analog) and DigiCipher II (digital) systems developed by General Instrument also share many common elements.

Digital encryption systems have their own equivalent of the PRBS, seed, and key. However, the implementation points of the digital encryption process are not limited to points along individual lines of video. Instead, the video implementation points are located within the individual 8 × 8 pixel blocks that comprise an MPEG-2 macroblock.

By way of analogy, the six blocks (four luminance and two chrominance) within a 4:2:0 macroblock can be compared to a six-sided puzzle introduced in the 1970s called "Rubik's Cube." With Rubik's Cube, the trick was to manipulate the puzzle until all squares of a given color were maneuvered to fill each one of the cube's six sides. Anyone who has attempted this feat knows that it usually took several days of diligent cube manipulation before the puzzle could be solved. Now imagine that midway to solving the puzzle someone comes along and randomly changes the colors for all the squares on your Rubik's Cube. Suddenly you are no closer to a solution than you were to begin with! This is just what happens when the encoder changes the electronic key and/or seed at periodic intervals.

All analog TV encryption systems are limited in their effectiveness because the techniques used are all spatial in nature: the video image is manipulated by rearranging the presentation order of video lines or line segments. The result is some sort of video image that appears on the TV screen. With digital TV encryption, however, manipulation of the quantization matrix results in a signal that the unauthorized IRD perceives as random noise. Rather than displaying a scrambled video image, the digital IRD will superimpose a message onto a black screen, informing the viewer that the IRD does not have the conditional access smart card required to decode the signal.

The methods used to encode each block's quantized matrix are proprietary and unique to each of the conditional access system providers. We can see, however, that all of these CA systems use mathematical algorithms to manipulate each quantized matrix in a more complex manner than analog TV encryption systems can achieve.

DIGITAL SYSTEM TROUBLESHOOTING

If one day the digital IRD stops receiving a picture, an installer or technician will need to use some form of signal tuning meter to determine the source of the problem. The digital IRD may come with a built-in signal meter with a readout that is displayed as an on-screen graphic. It is a good idea to note the signal level at the time of installation so that the technician can compare the working signal level to the level obtained upon installation if anything goes wrong with the system.

In addition to providing a signal level, some digital set-top boxes will identify the digital programming source as well as offer an audible tone indicator in a corner of the signal meter display, with the pitch of the tone rising as the signal rises. This is useful when the dish has been blown out of alignment by strong winds. The technician can turn up the volume of the TV set, then make fine adjustments to the dish alignment outside while listening to ascertain whether the audible tone indicator rises or falls.

Immediately following installation, it is also a good idea to put marks on the DTH antenna mount which indicate the correct antenna tilt (elevation) and corrected compass bearing (azimuth) angles for peaking the antenna. That way, the installer can confirm visually whether or not the antenna alignment has changed at some later date.

If the digital IRD does not have a built-in signal meter, the technician will need to connect some kind of outboard measurement device to the system. A small, lightweight, and inexpensive satellite signal meter is a practical device that technicians should have in their tool kits. This signal meter can be inserted into the coaxial cable that links the LNB or LNF and IRD. The same DC voltage that the IRD sends up the coaxial cable to power the LNB also provides power to the meter, so no battery or other power source is required to operate the instrument. One disadvantage of this type of signal meter is that it measures all of the signals coming from the satellite at once, so it is of little assistance in making polarization adjustments. The other main shortcoming of the low-cost signal meter is that its readout is not calibrated, so only a relative measurement of signal level can be obtained.

The spectrum analyzer is the single most powerful tool available to the professional satellite installer. It is a specialized receiver that continuously sweeps across an entire band of frequencies and presents the output as a video display of signal amplitude against frequency. The result is a panoramic view of the amplitudes and frequencies of all the signals present in the band of interest.

The portable, battery-operated spectrum analyzer displays signal amplitude from left to right and ascending frequencies from bottom to top. As the technician moves the dish through the satellite arc, he or she can detect signals visually and then peak the settings for each and every available signal. By adjusting the analyzer's frequency marker, the technician can display the center frequency for any one signal on the digital readout of the ana-

lyzer. The technician also can adjust the pass band of the spectrum analyzer to look at an entire satellite frequency band, or at just one satellite signal.

The shape of the signal displayed on the analyzer's screen also provides clues as to just what type of transmission is taking place. Conventional analog satellite TV signals have one kind of characteristic shape; digital signals, including compressed digital video signals, have another (Figure 6–12).

The spectrum analyzer is also the best tool for adjusting the separation between two orthogonal senses of polarization. Figure 6–13 shows several strong carriers with a few weaker ones sticking up from the noise floor to screen left. As the technician makes adjustments to the feed's polarization settings, he or she can verify visually that the system has been correctly adjusted to null out any unwanted cross-polarized signals.

Installers and technicians also can use the spectrum analyzer to measure the sidelobe performance of the dish. This will provide some indication of antenna surface accuracy and feedhorn positioning. All antennas generate sidelobes that amplify signals that are several degrees to either side of the antenna's main beam. The gain of these

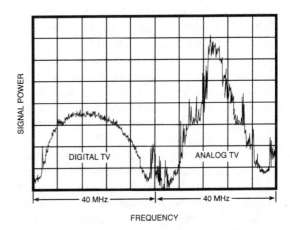

Figure 6–12 *Spectral comparison of digital and analog TV signals. (Courtesy of Taylor Howard.)*

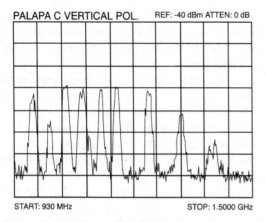

Figure 6–13 Wide-band spectral display of the vertical polarization C-band transponders on the Palapa C2 satellite. (Courtesy of Brett Graham.)

sidelobes must be –15 dB to –18 dB down from the main beam to prevent adjacent satellite signals from causing interference. The spectrum analyzer reveals the amount of sidelobe rejection, as well as how far off axis these sidelobes actually are.

The quantifying measure of an analog channel is the carrier-to-noise ratio, or *C/N*. A spectrum analyzer can be used to measure the *C/N* of the satellite receiving system. This *C* over *N* value represents the difference in decibels between the peak carrier and the average noise level hiding under the signal. To do this the technician first measures the carrier peak, and then jogs the antenna away from the satellite until only the noise floor can be measured. The technician also must make one mathematical correction to these numbers, which factors in the bandwidth of the satellite signal as well as the bandwidth of the filter built into the spectrum analyzer.

For example, the following measurement takes into account the performance characteristics of a popular portable spectrum analyzer model. As can be seen on the spectrum analyzer display, the carrier has an amplitude of –54 dB while the noise floor is –72 dB.

In the case of our spectrum analyzer model, the correction factor is equal to the satellite transponder bandwidth divided by the spectrum analyzer filter bandwidth × 1.5. (Other spectrum analyzers will have their own specific *C/N* measurement procedures, filter bandwidths, and recommended correction factors.) If the satellite transponder bandwidth is 36 MHz and the analyzer bandwidth filter is 8 MHz wide, then the correction factor would equal (36 divided by 8) times 1.55, or 6. 75.

The actual noise equals the noise floor measurement plus the correction factor of 6.75. (Actual noise: –72 dB + 6.75 = –65.25 dB; the *C/N* is the carrier minus the actual noise (*C/N*: –65.25 dB – (–54 dB) = 11.25 dB).

The quantifying measure of a digital channel is the bit energy to noise density ratio, or the E_b/N_o. Essentially, the received E_b/N_o represents the signal-to-noise ratio that the receiving system achieves. Another way to gauge the importance of E_b/N_o is to realize that as E_b/N_o increases, the number of bit errors decreases. Forward error correction is used to achieve a given BER at as small a level of E_b/N_o as possible.

The E_b/N_o is the carrier power C divided by the data rate f_b. Since data rate and E_b are associated, they must both either include or exclude the forward error correction overhead. As we previously have seen with the formula for the G/T, simple subtraction can be used to solve the equation when all values have already been converted to decibels.

$$E_b/N_o = C \text{ (dBm)} - N_o \text{ (dBm/Hz)} - 10\log f_b.$$

Terrestrial interference from land-based telephone transmitting stations, or airport, shipyard, and military radar installations, also can restrict or even preclude satellite reception. The spectrum analyzer displays the interference source when the LNB is pointed in the direction of our local microwave telephone relay station.

At some locations, the interfering signal consists of just a single narrow-band channel. In this case, the interference can be removed by using special narrow-band filters built into the receiver. In other cases, however, the interference may be much more pervasive, with strong multiple interfering carriers.

It is always a good idea to sweep any prospective site with a spectrum analyzer before performing an installation to obtain advance warning of any potential TI problems. The installer can then select the exact spot that will experience the least amount of interference. For example, rooftop installations are more susceptible to microwave interference than those on the ground.

Some installers elect to use a portable test antenna and spectrum analyzer to confirm signal strength and availability before installing the customer's dish. Others prefer to rely on the available satellite coverage maps to determine the satellite signal strength at the site location. Useful information on satellite performance can be obtained from the Internet web sites listed in Chapter 3.

INTERMITTENT RECEPTION PROBLEMS

Intermittent digital TV reception can be a tricky problem to solve. It can be caused by a variety of sources, such as the wind blowing the dish out of alignment, a loose feedhorn support bracket, or a loose cable connector that maintains poor contact or allows moisture to seep into the connection. A good starting point after detecting a loss of signal is to make a physical inspection of the entire system. Look for loose bolts on the antenna mount or feed and inspect all the cable connections. Check the reference marks on the antenna mount to ensure that the dish has not moved away from its proper alignment.

Rain, fog, or even rain-filled clouds rolling overhead can also reduce the intensity of Ku-band signals. During the installation process, it is tempting to stop making adjustments to the antenna and feedhorn as soon as perfect pictures appear on the TV screen. This is a major mistake. The picture will be crystal-clear when receiving a digital signal under clear sky conditions. Once the rain begins to fall, however, the picture will disappear.

Loss of signal acquisition during a rain outage appears in one of two ways. The digital IRD may display a freeze frame that represents the last video frame stored in the decoder's buffer circuit, while other digital IRDs will simply display a black screen with a "no signal" message superimposed on top of it. If reception frequently cuts in and out during a light rainstorm, this is a good indication that the system has not been peaked to maximum performance. The installer should use a signal-tuning meter to peak the antenna and feedhorn polarization alignments to ensure that you have the maximum amount of signal margin to counteract these environmental effects.

If the signal level indicator reads high but the system still does not receive any video, then the technician will need to ensure that no one has changed the basic IRD settings. Most IRDs have a parental control feature with a password setting that prevents unauthorized access to system settings and certain TV channels that the installer or operator can designate as off-limits. I know of one inquisitive 2-year-old who learned how to turn on adult TV movie services in a hotel suite while his parents were out of the living room. It is therefore advisable to use the IRD password feature to protect the indoor unit's high-tech settings from being changed, as well as to put limits on young inquisitive minds.

The most important settings for any digital IRD are the transponder center frequency and polarization, as well as the symbol rate and forward error correction (FEC) settings for the particular digital programming bouquet to which the customer subscribes. These settings usually are found under the "Installation" or "IRD Set Up" menu and are displayed as an on-screen graphic. Either the IRD will require the LNB IF frequency (usually between 950 and 2,050 MHz) or the LNB local oscillator frequency *and* the actual satellite frequency (for example: a local oscillator frequency of 5.150 GHz minus the satellite frequency of 4.000 GHz produces an IF frequency of 1,150 MHz).

The symbol rate and FEC settings often vary from one digital programming bouquet to the next. On the Astra satellite constellation, for example, symbol rates of 22 and 27.5 megasymbols per second (Msym/s) are in use. Other settings also are in use on other satellite systems around the world.

Check to be sure that none of these IRD settings have been tampered with. Some digital IRDs have an option that allows the restoration of the original factory settings at the touch of a button. If the system is now receiving a robust signal and all of the settings just mentioned are correct, the IRD will acquire the signal database and display the digital bouquet's electronic program guide (EPG). If the system still cannot receive a picture, check to ensure that the smart card is properly seated in the unit's conditional access slot and the subscription account is current with the program service provider.

RAIN FADE EFFECTS ON KU-BAND TRANSMISSIONS

There is one major drawback to satellites downlinking signals at frequencies greater than 10 GHz: the signal wavelength is so short that rain, snow, or even rain-filled clouds passing overhead can reduce the intensity of the incoming signals (Figure 6–14). At these higher frequencies, the lengths of the falling rain droplets are close to a resonant submultiple of the signal's wavelength; the droplets therefore are able to absorb and depolarize the microwaves as they pass through the Earth's atmosphere.

Digital DTH systems are designed to use a receiving antenna that is as small as possible, typically 60 to 75 cm in diameter. In places such as Southeast Asia (Figure 6–15) or the Caribbean, however, torrential downpours can, at times, attenuate the incoming Ku-band

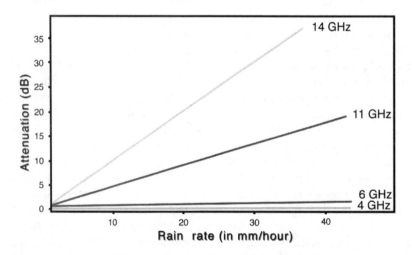

Figure 6–14 Attenuation rates in decibels for C-band and Ku-band satellite signals.

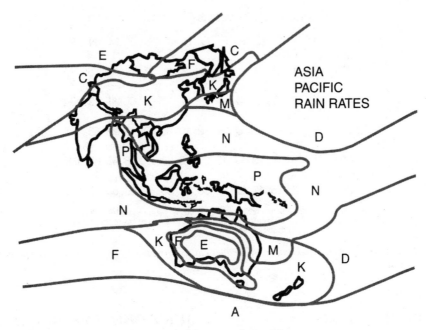

Figure 6–15 *Average rain rates (in mm/br) vary widely across the world.*

satellite signal by 20 dB or more; this may severely degrade the quality of the signals or even interrupt reception entirely. The duration of these rain fades, however, is usually very short and typically occurs in the afternoons or early evenings rather than during the prime-time evening viewing hours. For most Ku-band satellite TV viewers, these service interruptions will only amount to a few hours of viewing time lost over the course of any year. Moreover, consumers are used to experiencing outages from other utilities, such as their local power, telephone, and cable companies.

To help counteract the effects of rain fade, Ku-band system designers can use a larger antenna than would be required under clear-sky conditions. This increase in antenna aperture provides the system with several decibels of signal margin so that the receiving system will continue to function whenever light to moderate rainstorms occur. With digital DTH systems, however, consumer expecta-

tions compel bouquet operators to limit dish diameter to less than 1 m. For these systems, service interruptions are a given and the service provider typically will inform its customers to expect a certain percentage of outages per year.

In Malaysia, for example, DTH system operator Measat Broadcast Networks promises 99.7 percent signal availability, which is another way of saying that for an average of 0.3 percent of the time, service will be interrupted. Figure 6–16 shows average rain rates across the Asia/Pacific region; letters correspond to those on the map in Figure 6–15.

Satellite TV viewers in arid regions such as central Australia or the Middle East will rarely experience rain outages. In the Middle East, however, satellite dish owners may experience outages caused by intense sandstorms. The presence of any atmospheric particulate—even sand—can have an adverse effect on satellite TV reception.

Rain Rates in mm/hr versus Percentage of Time

Time	A	B	C	D	E	F	G	H	J	K	L	M	N	P
1.0	0	1	2	3	1	2	3	2	8	2	2	4	5	12
0.3	1	2	3	5	3	4	7	4	13	6	7	11	15	34
0.1	2	3	5	8	6	8	12	10	20	12	15	22	35	65
0.03	5	6	9	13	12	15	20	18	28	23	33	40	65	105
0.01	8	12	15	19	22	28	30	32	35	42	60	63	95	145

1 Year = 8,760 hrs. 0.3% = 26.28

99.7% availability in rain zone "K" requires that the receiver have sufficient margin above threshold to overcome rates of less than 6mm/hr.

Figure 6–16 *The amount of signal attenuation in decibels caused by rain can be expressed as the percentage of time within any given year.*

KEY TECHNICAL TERMS

The following key technical terms were presented in this chapter. If you do not know the meaning of any term presented below, refer back to the place in this chapter where it was presented or refer to the Glossary before performing the quick check exercises that follow.

Algorithm

Amplitude modulation (AM)

Conditional access (CA)

Electronic program guide (EPG)

Encryption

Frequency modulation (FM)

Integrated receiver/decoder (IRD)

Pseudorandom binary sequence (PRBS)

Rain fade

Seed

Smart card

Threshold

QUICK CHECK EXERCISES

Check your comprehension of the contents of this chapter by answering the following questions and comparing your answers to the self-study examination key that appears in the Appendix.

Part I: True or False

_____ 1. The conditional access component of an IRD is contained in each unit's smart card.

_____ 2. So-called "hard" encryption systems remove the horizontal and vertical sync pulses and invert the video to prevent unauthorized reception.

_____ 3. Adjustable bandwidth filters are an essential feature of any digital IRD.

_____ 4. The threshold of an analog IRD is the point at which the relationship between the incoming satellite signal's *C/N* and the *S/N* of the displayed video departs from a linear relationship.

_____ 5. Encryption of digital satellite TV signals takes place at the block level.

_____ 6. DVB-compliant IRDs are mutually interchangeable.

_____ 7. The "seed" element of an encryption system is transmitted over the satellite along with the program signal.

_____ 8. The IRD's pseudorandom binary sequence (PRBS) generator controls the implementation points for video encryption.

_____ 9. The digital IRD must be capable of tuning to the signal's exact transmission rate in megasymbols.

_____ 10. The threshold of a digital IRD is the point at which the relationship between bit error rate and MPEG-2 data rate is no longer linear.

Part II: Multiple Choice

11. Analog encryption system that cuts video lines and then repositions the individual segments:

a. sync suppression

b. line shuffling

c. line delay

d. line translation

e. conditional access

12. The numerical key that the encoder can frequently change at will is called the _____.

a. control algorithm

b. conditional access reader

c. seed

d. PRBS

e. none of the above

13. Each bouquet's digital signals consist of a unified bitstream, or _____.

a. uniform FEC rate

b. multiplex

c. QPSK envelope

d. conditional access waveform

e. service information graphic

INTERNET HYPERLINK REFERENCES

Digital Satellite System (DSS) IRD. Manuals for reception of DirecTV and USSB digital bouquets.
http://www.directv.com/manuals/index.html

General Instrument 4DTV IRD. Product information concerning reception of encrypted analog and digital TV services.
http://www.4dtv.com/brocspec/index.html

Nokia Multimedia Network Terminals.
http://www.nokia.com/products/multimedia/index.html

Scientific-Atlanta's PowerVu Digital Video Compression Networks. These are widely used to provide secure private business television networks to leading satellite TV programmers and broadcasters.
http://www.scientific-atlanta.com/D/satellitetvnetworks/products/index.htm

Chapter 7

Digital Satellite TV Installations

Digital satellite receiving systems are being installed around the world at an amazing rate that will only accelerate once digital satellites become the preferred method of providing high-speed wireless access to the World Wide Web. Many well-trained technicians will be needed to install all those dishes.

The difference between obtaining an excellent satellite TV signal and problematic reception often boils down to one thing: the installation. A correct installation performed by a professional installer can pull out that last fraction of a decibel in signal strength, making the difference between problematic TV reception and a perfect TV picture.

But what is it that makes an installation picture-perfect? It is not as simple as it may first appear. There are a lot of little things that can go wrong and numerous details to which the professional installer must pay attention. Moreover, the on-going evolution of satellite TV from traditional analog television digital video is placing new challenges before even the most experienced satellite technicians and installers.

This chapter provides a basic overview of the various stages required in the installation of satellite television receive only (TVRO) systems. The antenna and other major electronic components will come with their own assembly and operating instructions. Anyone who is relatively unfamiliar with the required installation procedures, however, will benefit from a concise overview of how the individual parts within the system function together, as well as how other criteria—such as site selection and initial antenna alignment—can affect the performance of the completed system.

CHECKING THE SITE

Satellite signals are microwaves that travel in a straight path along the line of sight. All geostationary satellites are located in an arc that goes across the sky. To receive any geostationary satellite, the site location must have an unobstructed view of that part of the sky in which the satellite is located.

From locations in the Northern Hemisphere, the satellite receiving antenna must have an unobstructed view of the southern sky, while in the Southern Hemisphere, the dish will need an unobstructed view of the northern sky. An inspection, or site survey, of the proposed earth station location should be made at the outset to ensure that there are no tall buildings, trees, power poles, or other obstructions that may block the signals from reaching the antenna.

Whenever a site survey is conducted during the autumn or winter, any nearby trees and foliage should be taken into account. They may end up blocking the antenna's view during the spring and summer months. Future construction plans of nearby buildings also must be considered; any new structures could be in a position to affect the reception.

The installer can conduct a preliminary site survey by standing at the proposed antenna site and looking to the south (or north for sites located below the equator). For total access to all available geostationary satellites viewable from the site location, a clear and unobstructed view of the sky all the way from the southeast to the southwest (or northeast to northwest for sites located below the equator) would be required. However, the view from many sites—especially those in metropolitan or suburban areas—will have obstructions in one or more directions. At these restricted sites, some satellites may not be viewable. To determine precisely which satellites will be visible from the site location, a working understanding of standard map coordinates as well as two site-specific angular coordinates is required.

LONGITUDE AND LATITUDE

Meridians, imaginary lines circling the Earth from pole to pole, cross over each of the Earth's 360 degrees. The distance from one meridian to any other is defined in degrees, minutes (1 degree = 60 minutes), and seconds (1 minute = 60 seconds) of longitude. The prime meridian, which crosses through the Greenwich Observatory in London, England, is referred to as zero degrees longitude and the two 180-degree segments to either side of the prime meridian are assigned ascending values of degrees east and west longitude. East and west meet again at the International Date Line that runs through the Pacific Ocean. References to any satellite's orbital location, as well as any receiving site location, are made in degrees of longitude.

The distance from the Earth's equator to any location north or south is defined in degrees, minutes, and seconds of latitude. The Earth's equator is referred to as zero degrees latitude and the two 90-degree segments to either side are assigned ascending values of degrees north and south latitude.

AZIMUTH AND ELEVATION

Azimuth and elevation are the two basic coordinates used to determine each satellite's position in the sky. The azimuth coordinate represents the bearing of the satellite from the site location, while the elevation is the angle at which the antenna tilts up toward the sky (Figure 7–1).

Every satellite within view of a specific installation has its own unique pair of azimuth and elevation coordinates. Given the site's latitude and longitude, the coordinates for any satellite can be determined through the use of a scientific calculator, a computer program, or available reference charts. An elevation/azimuth chart like the one in Figure 7–2 can be used to obtain the approximate azimuth and elevation coordinate values for most site locations.

Once the installer has determined the azimuth coordinate for a specific satellite, a compass can be used to find the precise

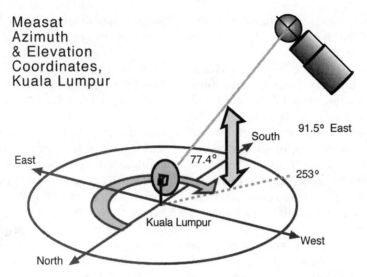

Figure 7–1 *Every geostationary satellite visible from the site location has its own unique pair of azimuth and elevation coordinates.*

direction from the site location. The initial compass reading, however, must be adjusted, or corrected, to account for variations between the magnetic North Pole that the compass senses and true North Pole upon which the Earth actually rotates. Although maps are available that can provide the required correction factor (Figures 7–3 and 7–4), the location of the Earth's magnetic North Pole actually shifts periodically. Unless the magnetic deviation map is fairly current, it is advisable to contact the control tower at the nearest airport to obtain the latest correction factor information for the area.

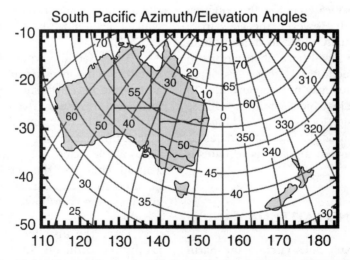

Figure 7–2 *Azimuth and elevation coordinates for 156 degrees east longitude in Australia and New Zealand.*

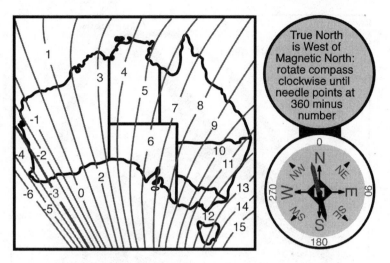

Figure 7–3 *Magnetic correction map for Australia.*

The correction factor must be subtracted from the compass readings whenever true north is east of magnetic north; it must be added to the compass readings whenever true north is west of magnetic north. Large metal objects or overhead AC power lines and transformers can affect the accuracy of compass readings. All measurements should therefore be made from a location that is well away from any manmade devices that can affect the accuracy of the compass readings.

At any given site location, the geostationary orbit describes an arc that runs from the eastern to the western horizons. The highest point of this arc, called the arc zenith (Figure 7–5), is located directly in line with the site longitude, which also is the true north/south line that transits the site. The greater the

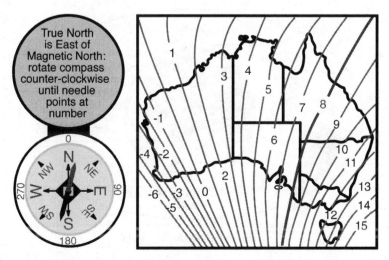

Figure 7–4 *Magnetic correction map for Australia.*

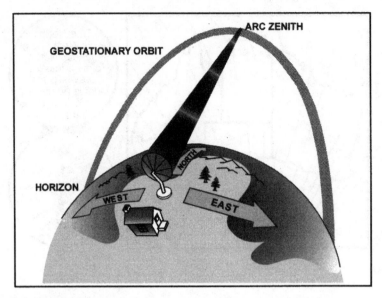

Figure 7–5 *Arc zenith is the location along the geostationary arc that is due south (locations north of the equator) or due north (locations south of the equator) from the installation site.*

difference between any satellite's azimuth coordinate and the site's true north/south line, the lower the value of the satellite's elevation coordinate.

Only a portion of the entire geostationary arc is visible from any one location on the Earth's surface (Figure 7–6). At some point, the elevation angle of satellites to the extreme west or east of the site location will fall below the horizon, obstructing reception. That is

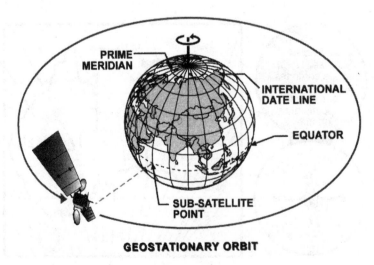

Figure 7–6 *Characteristics of the geostationary orbit.*

why satellite earth stations in Asia or the South Pacific cannot view North American satellites, and vice versa.

The total amount of the geostationary arc that is visible from any site location is a direct function of the site's latitude. Sites relatively close to the Earth's equator will be able to access a wider section of the geostationary arc than those located at the higher latitudes. For example, a site located at 15 degrees north latitude can potentially receive any geostationary satellite that is [±]76 degrees from the site's longitude. Another site located at 55 degrees north latitude will only be able to receive satellites that are [±]65 degrees from the site's longitude. The chart in Figure 7–7 allows the installer to quickly calculate the range of the visible arc for any site location.

An inclinometer is a standard satellite in-

stallation tool that measures the degrees of elevation, or "tilt," of the receiving antenna. The installer also may use an inclinometer to determine whether obstructions would block out any of the available geostationary satellites. Position the inclinometer on a level surface at the center of the site and align it to the corrected compass bearing, or azimuth, for the desired satellite. Then tilt the inclinometer back until its measurement gauge indicates the elevation angle of the desired satellite (Figure 7–8). By sighting along the inclinometer's straight edge, the installer can visually determine whether or not there are any obstacles between the proposed site location and the satellite's position in the sky.

If the installer initially encounters trees, foliage, or other obstructions that may interfere with reception, it may be possible to

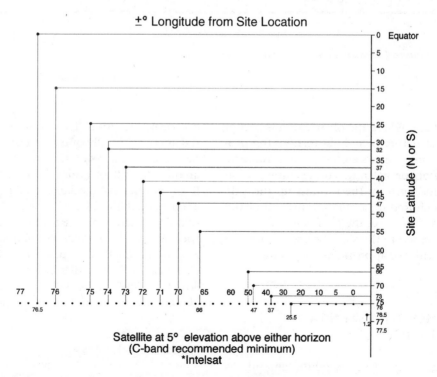

Figure 7–7 *Calculation of satellite visibility limits (a minimum of 5 degrees above the installation site's horizon is recommended for C-band reception and 10 degrees above the horizon for Ku-band reception).*

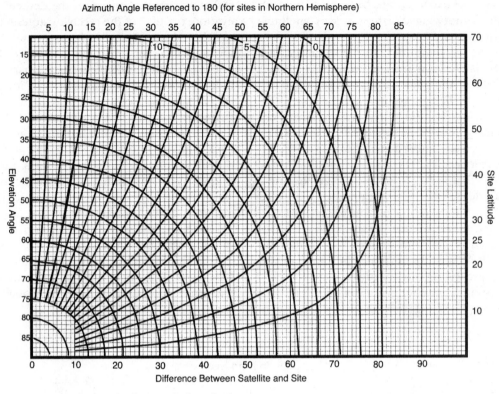

Figure 7–8 *Universal azimuth–elevation look angle chart.*

remove them. Where the obstruction is part of a nearby structure or natural terrain, the proposed site may be elevated from ground level to a rooftop or balcony. For example, a tall pole extending from the ground up the side of, and anchored to, a building may allow the antenna to "see" over the roof. Small antennas can be fastened to the side, eaves, or roof of a building with the appropriate hardware.

FIXED-MOUNT ANTENNA INSTALLATIONS

A small-aperture, fixed satellite antenna is most often used to receive a single digital DTH platform, or multiple satellites colocated at one orbital location. The mount supplied with the antenna usually offers a variety of possible placements such as on the ground, attached to an outside wall, or under the eaves of a house. Roof mounts are available that penetrate the surface of the roof and attach either to the building's rafters or directly on the top of a flat roof. Above all, the installer should select the type of mount that is appropriate for the precise location on site that has a clear and unobstructed view of the satellite or satellites of choice.

Installing the LNF

Most small dish systems provide a combination of feedhorn and LNB called an LNF—

short for low noise feed. Many of these products are designed to switch polarization from horizontal to vertical—or from right-hand to left-hand circular—by means of a DC switching voltage (usually 17/13 volts DC) that the IRD sends up the coaxial cable's center conductor. This simplifies the hookup between the outdoor and indoor units by limiting connections to a single coaxial cable. The LNF should be hand rotated while monitoring the signal level until peak performance is obtained from one sense of polarization. The IRD will switch the DC voltage to automatically reset the feed to receive the opposite sense of polarization whenever required.

High-power BSS satellite platforms, such as DBS-1, EchoStar I, and HISPASAT, use circular polarization to transmit digital DTH bouquets. In this case, there is no need to rotate the LNF in a clockwise or counterclockwise direction because it has been preset at the factory for optimum reception of circular polarization signals. Many other satellites, however, employ opposite senses of linear (horizontal/vertical) polarization. In this case, the LNF must be rotated or "skewed" in either a clockwise or counterclockwise direction to align it with the polarization of the incoming satellite signal.

The offset antenna most commonly used for digital DTH reception comes with a fixed-length LNF support bracket that sets the focal distance between the dish and feedhorn opening. There will, however, be some leeway for fine-tuning this distance—usually about an eighth of an inch in and out. The installer should manually make this adjustment while monitoring the signal level with a tuning meter or spectrum analyzer to ensure that peak signal performance has been obtained.

A length of coaxial cable will have to be installed between the location of the IRD inside the building and the LNF at the dish. Digital DTH systems that are equipped with multiple IRDs will require two lengths of coaxial cable. Cable clips should be used to fasten these cables to the side of the building. The

cable should be installed a bit below and then back up to the hole into the building, thus providing a "drip loop" to prevent rain from running down the cable and into the building. Type "F" connectors are crimped onto each end of the coaxial cable.

The power for the LNF also is supplied by this coaxial line. Whenever the satellite IRD is plugged into the building's AC wall receptacle, electricity is being sent up this line. To avoid a short circuit that would blow the fuse or circuit breaker in the IRD—or even damage the LNF—always unplug the receiver before connecting or disconnecting the system's coaxial cable.

Antenna Alignment

The alignment process is very simple and straightforward for DTH satellite TV systems dedicated to receiving a single satellite location. A mount for an antenna dedicated to receiving the Astra 1F and Astra 1G satellites colocated at 19.2 degrees east longitude, for example, will be adjusted to the site-specific azimuth and elevation coordinates, with the azimuth angle corrected from magnetic to "true" north for the site location.

Fine adjustments to the antenna's alignment will be made while receiving a digital satellite TV channel. When mounting the antenna onto its pole or mounting bracket, snug the bolts just enough to hold the antenna in place. Tightening the bolts down firmly on the pole should be one of the very last things done and should be completed only after the installer has "peaked" the antenna pointing for maximum signal level from a TV channel received from the desired satellite or colocated satellites. Tightening beforehand can dimple the pipe, making it more difficult to make subtle adjustments in alignment.

The antenna elevation angles supplied by computer programs, manuals, and other reference materials are almost always for prime focus antennas. When adjusting the elevation angle of an offset dish, subtract the offset angle that the manufacturer lists in the specifications from the elevation angle provided by the reference of choice (Figure 7–9).

Many offset dish manufacturers supply a gauge on the antenna mount that incorporates this correction (Figure 7–10). All too often, however, these gauges are not nearly as accurate as using an inclinometer to make a direct measurement (Figure 7–11). When using an inclinometer, the installer should place the tool onto any section of the mount that is in parallel with the rim of the dish and subtract the antenna's offset angle specification from the readings obtained. (The geometry associated with fixed dish alignment is shown in Figure 7–12.)

Installing the Digital IRD

Small-aperture offset-fed antennas most often are used to receive a digitally compressed bouquet of TV and audio services. With a conven-

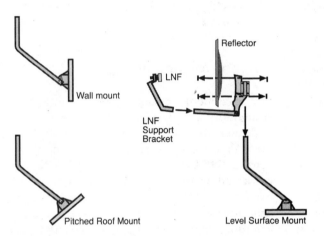

Figure 7–9 *Offset-fed antenna mounting options.*

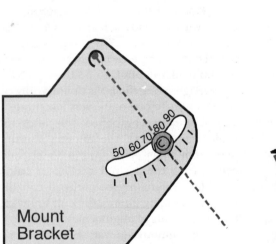

Figure 7–10 *Offset-fed antenna mount angle gauge.*

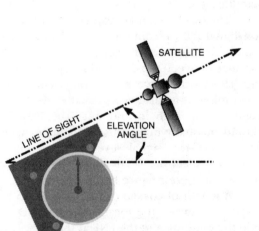

Figure 7–11 *An inclinometer can be used to determine line-of-sight access to any geostationary satellite visible from the installation site.*

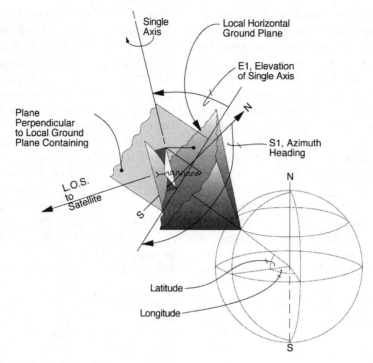

Figure 7–12 Single-axis fixed dish mount geometry.

tional analog TV signal any differences in picture fidelity that occur can be readily seen while the installer is tweaking the dish alignment. Under poor receiving conditions an analog TV signal is still visible. But the video is marred by "sparklies"—impulse noise that appears as comet-tailed dots streaking through the TV picture. But with digital, there either is a perfect picture or no picture at all. To verify that the highest possible signal level has been obtained, the installer will need to use some form of signal tuning meter to verify system performance.

Rain, fog, snow, or even rain-filled clouds rolling overhead can reduce the intensity of any Ku-band satellite signal. If the installer tunes a digital satellite TV system just by watching the TV screen, a crystal-clear picture may be obtained at the time of installation, but the customer may experience a signal outage as soon as rain begins to fall. Peak the signal reception to the maximum possible level to ensure that the customer has sufficient signal margin to overcome the attenuation caused by all but the highest rainfall rates.

Annual rain rates vary widely from region to region. Make sure that the customer is aware that some outages will occur. Some receivers will display a freeze frame picture during an outage that represents the last video frame stored in the decoder's buffer circuit. Other receivers will simply display a black screen. Tell the customer what to expect in order to avoid unnecessary service calls later on.

Many digital IRD manufacturers incorporate an audible tuning meter into their products. The installer can turn up the volume of the TV set and then go outside to make adjustments to the dish. Peak signal acquisition occurs when the tone rises to its highest level. As the antenna is adjusted slightly to the right/left and up/down, the audio tone provides an immediate indication of any changes in signal intensity (see Figure 7–13).

An in-line, signal-strength meter is another tool that can be used for performing this task. What's more, a dedicated signal-strength meter also will be more accurate and sensitive than the meter built into the satellite TV receiver. Some portable meters also will provide an audible tone indicator. When using a signal strength meter, take care to ensure that the correct satellite is being viewed before beginning the peaking procedure. Scan through the available channels at the outset—and again at the end—of the installation process to ensure that acquisition of the correct satellite has occurred.

Some signal strength meters will supply the voltage needed to power the LNF. They

therefore can be connected directly to the LNF IF output connector. Signal meters that do not supply power to the LNF will have to be inserted in the coaxial line running from the receiver to the LNF using a device called a three-way "tap". This tap supplies three connectors: one connects to the LNF, the second connects to the digital IRD, and the third supplies the connection to the meter so that the signal strength measurements can be made. Tighten down all of the mounting bolts as soon as the antenna alignment procedure has been completed. This will prevent strong winds or rain from repointing the antenna.

Manufacturers of digital IRDs typically program default values into each unit at the factory that correspond to the frequency, polarization, symbol rate, and forward error correction (FEC) rate of a single digital DTH bouquet. In the case of digital subscription services, the IRD also is preprogrammed to be fully compatible with the specific conditional access format that the digital DTH service provider is using to control subscription authorization. Systems manufactured specifically to receive a single digital DTH bouquet therefore will not need any adjustments to these factory settings. To receive a bouquet that uses different settings, however, new values must be programmed into the IRD. Instructions on how to reprogram these transmission parameters are provided later in this chapter.

LARGE-APERTURE DISH INSTALLATIONS

If assembly is conducted at ground level, several people will be needed to lift the reflector onto the antenna mount; alternatively, a crane or other mechanical device can be used to hoist the antenna into place. Other large-aperture antennas are designed so that they can be assembled, piece by piece, directly onto the antenna mount itself. All of these construction methods require that the installer check the curvature accuracy of the antenna following

Figure 7–13 *The installer should manually adjust the LNF in or out while monitoring the signal level with a tuning meter to ensure that peak signal performance has been obtained.*

assembly to ensure that the surface has retained a true parabolic shape.

For most large C-band installations, the installer will need to construct a concrete foundation for the mount. In the case of a pole-mount antenna, a steel pipe is placed in a hole in the ground and embedded in concrete. Because it will take a few hours for the concrete to harden, the pole should be set the day before construction is to begin. Quick-drying concrete products are available that will "cure" in less than an hour. Be sure to use a level to ensure that the pipe is "plumb" before the concrete hardens.

Other types of dish mounts will require a concrete pad for the foundation. Concrete piers extending well below the frost line should be incorporated into the pad design. Certain types of antenna mounts merely need a level surface.

Antenna Assembly

To obtain the maximum performance from any satellite dish, the antenna surface must conform to the parabolic curve selected by the engineer who designed the antenna. Any noticeable bumps or waves along the surface can affect the level of the signal arriving at the focal point of the dish. The installer can examine antenna accuracy by running his or her hand along the antenna surface and feeling for any noticeable bumps or waves. A close visual inspection will also reveal whether or not the screen for a mesh antenna has been properly attached.

Many mesh antennas use a series of clips to hold the mesh onto the antenna's frame. These clips must be attached to the frame at regular intervals in such a way that there are no bumps or waves along the surface. Care taken by the installer when putting the mesh onto the frame will be rewarded with the best reception possible.

Most solid metal and fiberglass antennas are constructed from a series of petals that fit together to make the dish (Figure 7–14). The installer should take care to ensure that there is little variation from petal to petal.

Another way to check the surface accuracy of an antenna is to sight across the rim from one edge to the other. The near and far edges should line up parallel to each other. If

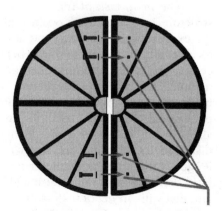
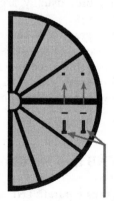

(4) 1/4"-20x2"Bolts
(8) Flat Washers
(4) 1/4"-20 Nuts

(2) 1/4"-20x2"Bolts
(4) Flat Washers
(2) 1/4"-20 Nuts

Figure 7–14 *Assembly of a petalized mesh antenna.*

they do not, the installer may have to loosen some of the bolts and retighten them in a way that favorably alters the shape of the reflector.

System Noise Performance

Molecular motion within all matter generates a noise background that permeates the entire electromagnetic spectrum used to propagate communication signals, including the satellite frequency bands. The temperature of all thermal noise is expressed on the Kelvin scale, which can be related to other more familiar temperature scales such as Celsius or Fahrenheit. The higher the noise temperature, the stronger the noise sources.

The satellite antenna receives externally generated noise along with the desired signal. When the satellite dish tilts up towards the "cold" sky, its noise temperature is relatively low. If the antenna must tilt downward to receive a low-elevation satellite, the noise temperature will increase in a dramatic way because it is now able to see the "hot" noise temperature of the Earth. The actual amount of noise increase in this case is a function of antenna diameter. Minimum antenna elevation angles of 5 degrees, for C-band, and 10 degrees for Ku-band, usually are recommended.

The noise temperature of the receiving system will also increase if moisture is allowed to enter any of the system's connections. In some cases, moisture can cause irreparable damage if it is allowed to corrode sensitive internal circuitry.

Mounting the LNB and Feedhorn

When bolting together separate LNB and feedhorn units, the installer should take extreme care not to touch the short metal probe inside the mouth of the LNB. Any grease or dirt that adheres to the LNB probe's surface will degrade the unit's performance. There is a neo-prene gasket that is inserted between the feedhorn flange and the mouth of the LNB. Unless this gasket fits snugly in the groove provided, moisture will seep through this opening and collect in the mouth of the LNB where it can disrupt the reception or even cause damage. The LNF, which comes as a single, sealed unit, is not subject to these problems.

The installer should use and snugly tighten each and every one of the bolts provided by the feedhorn manufacturer. In tropical and semitropical climates, the installer also should seal the outer portion of this flange with a sealing compound such as silicon sealer. Otherwise, moisture may seep around the neoprene gasket and enter the mouth of the LNB.

Check to ensure that the feedhorn is centered over the dish. You can do this by taking four measurements at equidistant points from the feedhorn to the rim. All of these measurements should be equal. Another measurement that will vary from antenna to antenna is the focal distance—the distance from the center of the dish to the lip of the feedhorn. This is specified in the antenna installation manual. The angle of the face of the feedhorn also should be the same as the back plate of the dish. Use an inclinometer to check that this is the case.

The main task of the feedhorn support structure is to accurately position the feed at the exact center of the dish and at the correct focal length, that is, the distance between dish center and the feedhorn opening that the manufacturer recommends (Figure 7–15). This information is supplied in the antenna assembly manual. The focal length for any antenna also can be computed if the diameter of the dish and its f/D ratio are known; the focal length is the antenna diameter times the f/D ratio. For example, the focal length of a 10-ft antenna with an f/D of 0.45 = 10 x 0.45 = 4.5 feet (54 inches). Once the LNB and feedhorn have been installed on the support structure, the installer can make incremental adjustments to the focal length while monitoring a signal strength meter.

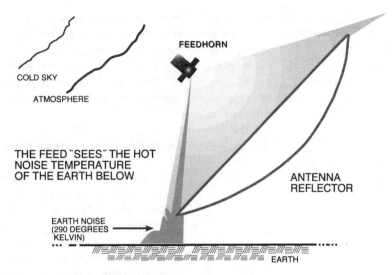

FEEDHORN

COLD SKY

ATMOSPHERE

THE FEED "SEES" THE HOT
NOISE TEMPERATURE
OF THE EARTH BELOW

ANTENNA
REFLECTOR

EARTH NOISE
(290 DEGREES
KELVIN)

EARTH

Figure 7–15 Incorrect adjustment of the feedhorn can result in overillumination of the dish, where the feed sees over the rim of the antenna and intercepts thermal noise from the "hot" Earth below.

To determine the antenna diameter, measure across the surface of the dish from one side to the other. The radius equals one-half the antenna diameter. The depth of the dish is the distance from the center of the dish to the plane of the rim. Stretch a string across the antenna's rim so that it crosses in the center of the dish. The depth will be the distance from the antenna's center to the string.

The prime focus antenna typically uses one of two support methods. The "buttonhook" support consists of a single piece of tubing that extends outward from the center of the dish to the focal point of the antenna. It is called a buttonhook because the curved support tube resembles a hook. The main advantage of this design approach is that the distance between feedhorn and dish can be incrementally adjusted by sliding the buttonhook in or out in relation to the center of the dish.

The main drawback of the buttonhook design is instability, especially when multiple LNB/feedhorn units are mounted at the focal point. The heavy weight of the feedhorn/LNB assembly causes the feed to shift positions whenever the antenna moves or strong winds

and rain push against the antenna reflector. Guy-wire kits are available that provide the additional stability required for complex installations that may have more than one LNB. Because of their narrow beam widths, Ku-band satellite antennas usually do not employ a buttonhook feedhorn support.

Other antenna designs use multiple support legs to improve the stability of the feedhorn and LNB (or multiple LNB units for dual-band systems). The number of support members are typically three ("tripod") or four ("quad") straight, equal length pieces of aluminum or steel that attach to equally space locations, either at the antenna's rim or on the antenna surface. Quad supports offer enhanced stability over the button hook design approach and therefore are recommended for installations equipped with more than one LNB.

The Scalar Feedhorn

Prime focus receive-only antennas usually require a feedhorn that can be broken down into two parts: a round flat "scalar" plate with concentric rings and a "waveguide" onto which the LNB is mounted. This waveguide (technically referred to as an "orthomode coupler") fits into the center of the scalar plate and can be adjusted inward and outward.

The distance that the waveguide extends beyond the surface of the scalar plate must be set to correspond to the *f/D* ratio of the antenna (Figures 7–16 and 7–17). The waveguide may be marked to indicate the various *f/D* ratio settings. Alternatively, the feedhorn

may come with an adjustment gauge for setting the correct location of the scalar rings. Consult the manufacturer's assembly directions or use the formula previously provided to determine the correct *f/D* ratio for the prime focus antenna being installed.

Installing the Direct Burial Cable

Steerable large-dish installations most always require a minimum of three groups of wires to connect the indoor and outdoor electronics. All of these wires typically are contained in a single direct burial cable (Figures 7–18 and 7–19). A shielded wire cable called coax is used to connect the LNB to the IRD. Coax is made up of an inner wire covered with a plastic or foam sheath, and an outer mesh that is in turn surrounded by an outer plastic covering. When installing connectors at each end of the coaxial cable, the installer should take care that the cable's center conductor does not touch the outer ground sheath, as this short circuit will blow the fuse inside the IRD and potentially damage its internal power supply. Special right-angle connectors are available that the installer can use whenever space limitations prevent a straight-on connection.

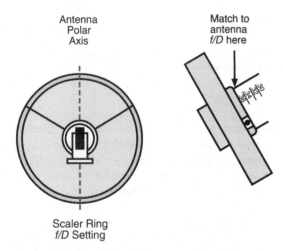

Figure 7–16 Scalar feedhorns require that the installer adjust the scalar plate to the mark that matches the f/D *ratio of the receiving antenna.*

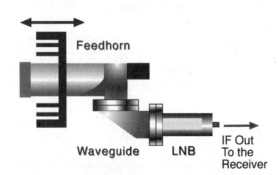

Figure 7–17 Close-up view of scalar feedhorn components.

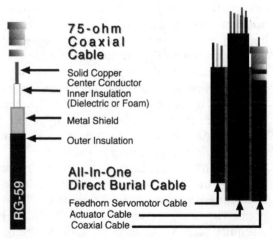

Figure 7–18 All-in-one direct burial cable connections.

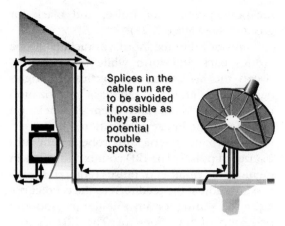

Figure 7–19 *Planning the cable run.*

A special type of cable TV connector, called an "F" connector, is crimped onto each end of the coaxial cable. (A few of the digital IRD products manufactured in Europe alternatively use an "IEC" coax connector instead of the F connector.) This connector mates with complementary connectors on the LNB and indoor receiver. A special tool is used to crimp "F" connectors onto the cable. Standard

lengths of cable with the connectors already installed also can be purchased at local supply houses.

Once satisfied with the total system operation the installer will need to go back and waterproof the connection out at the LNB by wrapping it with a sticky waterproof compound, such as Coax-Seal. For locations with a high incidence of rainfall, the installer should unplug the IRD, flood the inside of the coaxial connector with silicon seal, and then seal the outer portion of the connector with Coax-Seal. Do not plug the IRD back in until the silicon sealer has had ample time to dry or you will short out the receiver's fuse.

Satellite TV installations use a special type of coaxial cable that has a characteristic impedance of 75 ohms. CB radios and other two-way radio equipment use another type of cable with a characteristic impedance rating of 50 ohms that is not suitable for satellite TV use. Several different types of 75-ohm coaxial cable are available (Figure 7–20). RG-59U coax can be used to span distances of up to 100 feet. For longer lengths, lower loss RG-6 or RG-11 is used. All-in-one satellite

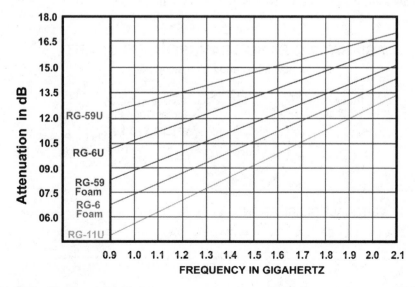

Figure 7–20 *Coaxial cable attenuation chart.*

cable contains one or two spans of RG-6. Since RG-6 is slightly larger in diameter than RG-59, it also requires a slightly larger F connector. To span distances of several hundred feet, special UHF line amplifiers with +10 or +20 dB gain also may be necessary to compensate for the amount of signal loss or attenuation that occurs as the signal passes along the length of cable.

Each line amplifier receives its power from the center conductor of the coaxial cable. Lockable, above-ground enclosures called "pedestals" are recommended to protect the amplifier from the environment and permit easy servicing should the amplifier ever malfunction.

The installer should unplug the IRD from the AC wall receptacle before connecting or disconnecting the coaxial cable from either the indoor or outdoor electronics. This eliminates the chance of accidentally causing a short circuit by touching the center conductor of the coax to ground.

The feedhorn cable consists of three shielded 22-gauge (or larger) stranded wires that connect the power, pulse, and ground connections of the IRD to the corresponding input terminals on the feedhorn servomotor. These wires also are color-coded (usually red

for power, white for pulse, and black for ground; see Figure 7–21).

Inside the feedhorn, a pickup probe swings back and forth while switching between the horizontal and vertical (or right-hand and left-hand circular) polarization transponders (odd and even channels on satellites using linear polarization). As the servomotor rotates the pickup probe, it generates a series of pulses. The IRD counts and stores in memory the number of pulses required to successfully adjust the feedhorn to the correct polarization setting for any satellite transponder or group of transponders. The IRD disconnects DC power to the servomotor as soon as the required pulse count has been generated to achieve the optimum polarization angle.

Keep in mind that there are built-in limits to the probe's clockwise and counterclockwise movements. The installer will need to align the feedhorn with respect to the antenna's polar axis so that the probe can rotate the 90 degrees from horizontal to vertical (or left-hand to right-hand circular) polarization without reaching the limits of its travel. Several manufacturers include a plastic directional guide with their products that visually indicates the proper alignment of the feed when installed on the dish.

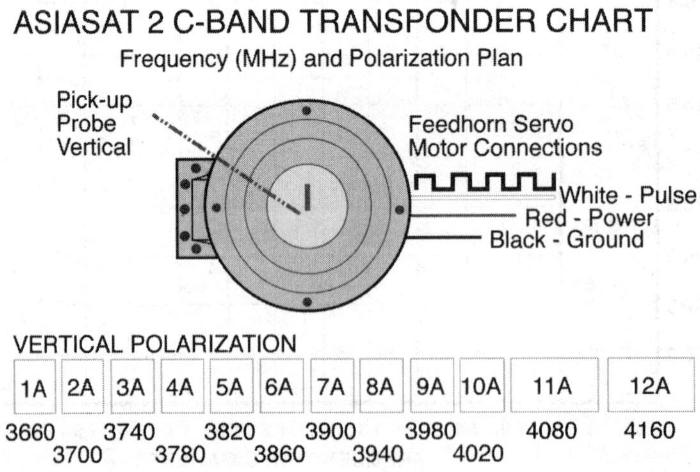

Figure 7–21 Scalar feedhorn wiring diagram.

If the feedhorn is not installed in the correct orientation, the pickup probe will run into its limits of travel before it can move beyond peak signal level. In this case, the installer should loosen the clamp that holds the feedhorn onto its support bracket and physically rotate the feedhorn in either the clockwise or counterclockwise direction until it is possible to rotate the feedhorn's pickup probe to pass the peak level for each sense of polarization.

The direct-burial cable's actuator line consists of five stranded wires. The wires should run through the actuator's rubber grommet, which seals off the opening so that moisture cannot enter the actuator housing. The installer also should mount the actuator motor so that the housing's drain holes are on the bottom to provide an exit point for condensation.

Two of these are 14- or 16-gauge stranded wires that connect to the large wire terminals at the actuator motor and to the motor "1" and "2" terminals on the back of the IRD. Three 22-gauge shielded wires, which connect to a sensor that is built into the actuator motor's housing, provide the required power, pulse, and ground connections to the IRD. Most actuator sensors require pulse and ground connections but not DC power. Consult the actuator manual to determine if this is the case.

Once all the connections have been made, the installer should move the dish to the east or west. If the antenna moves in the opposite direction to the one intended, reverse the wires connected to the motor wire "1" and "2" terminals.

Lightning Protection

It is extremely important that the satellite antenna have a proper electrical ground to prevent lightning strikes from damaging the outdoor electronics or gaining entry into the home. If the home's AC electrical ground or

"earth" is in close proximity to the antenna, the installer can use a No. 10 AWG or larger solid copper ground wire to connect the antenna mount to ground.

For those sites where the antenna is physically removed from the house, a separate grounding rod should be installed next to the dish. Use a No. 10 AWG or larger solid copper ground wire to connect the mount to copper rod. For additional protection, an antenna discharge unit should be connected in-line between the outdoor and indoor units to shunt transient voltages to ground before they can gain entry to the home. This device also must have an electrical connection to ground, either to a separate ground rod at the cable entry point to the home or to the building's AC ground. The installer also should put surge protectors on the AC receptacle that supplies power to the IRD and any other AC-powered devices that connect to the satellite system, such as a TV set, VCR, or home entertainment system.

The installer should suggest additional precautions to owners that will protect their systems from lightning strikes and power surges. Whenever there is a distinct possibility of a thunderstorm occurring, the system owner can unplug the IRD from the AC wall outlet and then disconnect the incoming coaxial cable or cables from the indoor unit. Once the storm front has passed, the owner can reconnect the coaxial cable to the IRD IF input port and then plug the IRD back into the AC wall outlet.

The Modified Polar Mount

The classical polar mount used by astronomers has an axis that is parallel to the polar axis of the Earth. DTH systems, however, must use a modified version of classical polar mount geometry that incorporates an offset angle that compensates for the relatively close proximity of the satellites (when compared to distant star systems and galaxies). This "declination" angle tilts the antenna slightly downward toward the plane of the Earth's equator to permit antenna steering in "hour angle" (i.e., along the geostationary arc) by rotation about a single axis (Figure 7–22).

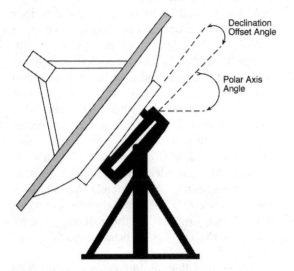

Figure 7–22 Alignment of the modified polar mount. The declination offset angle tilts the antenna's reflector downward in the direction of the geostationary arc.

To properly track the geostationary arc, the mount's polar axis must be accurately aligned to the true north/south line that runs through the site location. Any inaccuracies in this alignment will cause uneven tracking of the geostationary satellite arc (Figure 7–23).

It simply is not possible to obtain the required accuracy from corrected compass readings alone. At the outset, the installer should snug the bolts that secure the mount to its supporting pole or tripod just enough to hold the antenna in place. These bolts should be tightened down firmly only after the installer has used a signal meter or other instrument to confirm that the antenna is accurately tracking all of the available satellites.

The mount's polar axis and declination angles must be set precisely if the antenna is to accurately track all of the available satellites. To set the polar axis angle, place the inclinometer onto the mount's polar axis. Adjust the polar axis to the correct elevation angle for the site location.

The declination angle is a function of the latitude of the site location. Regional charts are provided later in this chapter to allow the

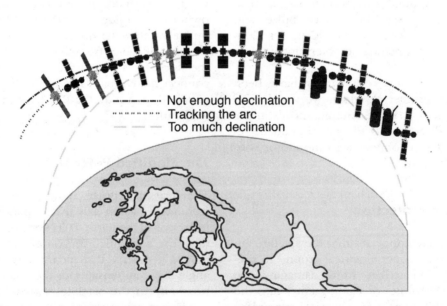

Figure 7–23 An incorrect declination angle results in poor tracking of the arc.

installer to determine the elevation and declination angles required for any site location. To set the declination angle, place the inclinometer on the back plate of the antenna (same plane as the rim of the dish). Adjust the declination so that the antenna elevation angle is equal to the polar axis angle plus the number of degrees of declination required for the site location. The dish should tilt down slightly from the mount's polar axis angle.

Installing Actuator Motors

Mounts equipped with a retractable actuator arm are best suited for locations where the desired satellites are located either on the eastern or western side of the site's true north/south line. The maximum length of a fully extended actuator arm typically is 18–24 inches for antennas that are less than 12 ft (3.7 m) in diameter. This gives the mount access to all satellites within a 70-degree range in longitude. Longer actuator arms are available for larger antennas.

For sites located in the Northern Hemisphere, the arm attaches on the western side of the dish for systems located in the eastern portion of the continent and on the eastern side of the dish for systems located in the western portion of the continent. For locations in the Southern Hemisphere, the actuator arm attaches on the eastern side of the dish for systems located in the eastern portion of the continent or on the western side of the dish on systems located in the western portion of the continent. If the site location is somewhere in the middle of the continent, the installer should observe how existing actuator arms are mounted on systems that already have been installed in the area.

The actuator arm will be fully retracted when it is first installed. Before tightening the clamp that holds the actuator arm in place, the installer should place the inclinometer on the back plate of the dish and take a reading. The elevation angle of the antenna should be slightly below the site-specific elevation angle for the satellite that is closest to the horizon. If an insufficient amount of room has been left to allow the antenna to move past the last desired satellite at either end of the arm's limits of travel, the clamp on the actuator arm can be loosened and adjusted slightly to permit further travel.

The heavier and more expensive horizon-to-horizon mount is best suited for those locations where the satellites of interest are to be found on both the eastern and western sides of the site's true north south line. The "horizon" mount, which can access all the satellites that are visible from the site location, typically has a 150-degree range in longitude. Instead of interfacing with a retractable arm, the motor attaches directly to a mating flange on the mount's gear reduction box.

Programming the East and West Limits

At this stage of the installation, the IRD should connect directly to the TV and not through any other component, such as a VCR, video switcher, or RF splitter. The appropriate IRD output connector for DTH systems will be labeled "To TV" or "RF OUT." On most units, the output signal can be switched between two or more VHF (or UHF for areas of the world where this is the predominant TV frequency range) TV channels. Select a channel that is not in use locally and tune the TV set to receive it.

The instructions for setting the antenna's programmable limits are the same for either the actuator-arm or the horizon type of mount. The IRD will prompt the installer with instructions that are displayed on the TV screen or on the front panel of the IRD. The first instruction most often encountered is to "Set East and West Limits," that is, the physical limits of travel for the actuator arm or horizon drive.

The motor has a slip clutch to prevent damage when any actuator arm has extended or retracted completely. The installer, however, should set the programmable limits so that the dish can move past the last satellite at either end of the visible geostationary arc but stop before engaging the motor's slip clutch.

If an actuator arm does reach its physical limit, the slip clutch will begin making a loud clicking noise. Stop immediately. If the dish becomes stuck in this position, take the motor off the actuator arm and insert the blade of a heavy screwdriver in the slot where the motor normally engages the arm. Turn the screwdriver just enough to loosen the arm, then put the motor back in place.

To locate satellites in the lower section of the geostationary arc, manually push the antenna right and left on the pole while moving the drive east and west in slight increments until the strongest signal levels have been obtained. Then snug the mount bolts onto the pole so that the dish will no longer rotate. Do not tighten them down firmly yet, however. Note the numerical reading that indicates the antenna's relative position. This number, which is displayed on the front panel of the IRD or on the TV screen, changes as the dish moves.

Move the dish east or west until it arrives at arc zenith. Adjust the digital IRD to the parameters for a digital bouquet on the satellite that is closest to arc zenith. Move the antenna in the direction of this satellite until the audible tone from the IRD tuning meter indicates that the digital signal has been acquired.

To locate satellites in the upper section of the geostationary arc, make incremental, up-and-down adjustments to the mount's elevation bracket and then jog the actuator drive east and west. Stop as soon as you have determined the position of maximum signal acquisition. The rule of thumb here is to adjust the elevation and use the actuator to receive the satellites in the upper portion of the arc. Do not adjust the elevation to receive satellites in the lower portion of the arc.

Fine-tune tracking by repeating these steps until the satellites in both the upper and lower sections of the geostationary arc achieve at their maximum signal strength without requiring any further adjustments. If this cannot be done, a miscalculation in the elevation or declination settings probably has been made.

Once the antenna is tracking properly, firmly snug the bolts that secure the mount onto the pipe. Placing a bolt all the way through the pipe may be desirable in high-wind areas. Mark the pole and the mount for later reference.

The final step in the antenna installation involves locating the correct antenna position for each of the available satellites and programming these positions into the memory circuitry of the digital IRD. When moving the dish to the east or west, the antenna look angle will now automatically adjust itself to track the geostationary arc.

Digital IRD Setup Procedures

All digital DTH services use a form of modulation called QPSK, short for quadrature phase shift keying. With QPSK, two data bits are combined to produce a representative symbol that is expressed as one of four distinct states or phases. This phase shift typically occurs at rates of millions of symbols, or megasymbols, per second (Msym/sec).

Symbol rates may vary widely from one digital DTH service to the next. It is important that the digital IRD be able to tune through the entire range of symbol rates that correspond to the services the customer wishes to receive.

The FEC rate is the ratio of k bits entering the uplink encoder to n bits exiting the encoder. FEC rates of 1 to 2 (1/2), 2 to 3 (2/3), 3 to 4 (3/4), 5 to 6 (5/6), and 7 to 8 (7/8) are now in use worldwide. The uplink encoder inserts error-correcting bits into the original signal's digital bitstream. The IRD uses this extra information to detect and compensate for signaling errors that may occur in the link between program source and destination.

With some digital IRDs, the LNF local oscillator (LO) frequency must be entered into the unit's setup menu along with the digital bouquet's default transponder frequency. The LO frequency, which is printed on the LNF product label, is typically 5,150 MHz for C-band LNB units. Keep in mind, however, that a few LNB manufacturers use other LO values.

To switch between digital DTH bouquets on the same satellite, enter the new transponder frequency and polarization, change symbol rates, and switch to the new FEC setting. Keep in mind that it will take a few seconds for the IRD to load the new data. Select the manufacturer's default settings to go back to the original bouquet.

Tune the IRD to the transmission parameters for a known digital DTH bouquet. If necessary, refer back to Chapter 3 to locate the correct frequency, polarization, symbol rate, and FEC rate. Engage the signal tuning meter and turn up the volume on the TV set so that you can hear the audible tone while standing outside at the antenna.

Auxiliary IRD Connections

A coaxial cable can be connected directly from the "To TV" ("RF Out") output of the digital IRD to the "Antenna" input of the TV set. The local antenna cable, which formerly was connected directly to the TV set, should now be connected to the "Antenna" input on the back of the digital IRD. To view local channels, turn the IRD off and the local channels will automatically appear on their respective channels on the TV.

If the IRD does not have this feature, install an A/B switch from a local satellite or electronics store. Both the IRD and local antenna connect to the "A" and "B" input ports on this switch, while the single output port connects to the antenna input on the TV. This switch will have to be changed manually. An external A/B switch also may be necessary if a second TV is connected to the IRD and one person wants to

watch local TV while the other is watching a satellite TV program.

There are other options when it comes to connecting the satellite system to any TV set or home theater system. The IRD supplies video and stereo audio outputs on its back panel that can be connected directly to a TV monitor or to a VCR for recording programs. Whenever a VCR is part of the overall entertainment system, the output of the VCR must be connected to the TV set and the local antenna connected to the "Antenna" input of the VCR (Figure 7–24). The stereo audio outputs on the back panel of the IRD can be connected directly to the stereo inputs of the TV set or to the auxiliary inputs of a home entertainment system's audio amplifier or tuner.

THE SCART CONNECTOR

Developed in France, the SCART (for Syndicat des Constructeurs d'Appareils Radio Recepteurs et Televisieurs) connector was developed in part as a standard for making connections between various audio and video products in the home entertainment system. The SCART connector, also known as the Peritel or Euro-Connector, subsequently was adopted as the standard audio/video interface connector by most countries in Western Europe.

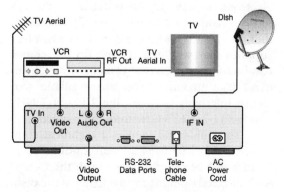

***Figure 7–24** Digital IRD wiring diagram.*

One or more SCART connectors are now found on all satellite receivers, IRDs, and decoders manufactured for the European market. Moreover, all TV sets sold in the European market have one, two, or even three SCART connectors! The SCART connection is advantageous for displaying video of the best possible quality on top-of-the-line TV sets. The TV output of the IRD is available in three modes. The RF output connects to the UHF antenna jack on a TV set or VCR. The composite video output connects to the video input or SCART connector of a VCR or TV/monitor. The RGB (Red-Green-Blue plus synch) connects to the R, G, and B circuits within a TV monitor.

The SCART connector can supply both composite video and RGB plus synch. Since a direct RGB connection can made between the satellite TV receiver, VCR, and TV, the visual imperfections associated with composite video are eliminated. SCART control centers and splitters with stereo audio output connectors also are available to enable TV audio sound to be played through any home entertainment system.

DUAL-BAND RECEIVING SYSTEMS

In many part of the world, dual-band satellite TV installations provide the best of the C- and Ku-band worlds. Special dual-band feedhorns are available that place both the C- and Ku-band feed openings at the focus of the dish. An electronic switch can be installed at the antenna to connect the main coaxial cable to either the C-band, or Ku-band feed. The receiver supplies the switching voltage up the center conductor of the coaxial cable. Universal Ku-band LNB products also are available that switch internally between the 10.7–11.7 GHz and the 11.7–12.75 GHz frequency spectra.

The dual-band feed is a technical compromise that sacrifices a small amount of C-band signal in order to place both the C- and Ku-band feed openings at the focus of the dish.

For those situations where the installer cannot afford to sacrifice any C-band signal, a second Ku-band feed and LNB can be attached to one side of the C-band feed along the plane of the dish that is perpendicular to the mount's polar axis. Another option is to install a second Ku-band dish that is permanently pointed at a single satellite, or satellites colocated at a single orbital position (Figure 7–25).

Multiple Feedhorn Installations

For SMATV systems, the installer may elect to install a multiple feedhorn array so that a single prime focus antenna can simultaneously receive signals from satellites that are adjacent to the central satellite at which the antenna points. For this method to work, the receiving antenna should be relatively shallow, that is, the dish should have a focal-length to antenna-diameter (f/D) ratio of 0.35–0.45.

The multiple feedhorn array along with the associated LNB units (C-band, Ku-band, or both) are aligned in a plane that is perpendicular to the polar axis of the prime focus antenna. The mounting bracket used to support the multiple feedhorn array is constructed in a such way that the installer can adjust the positions of the secondary feeds so that maximum signal levels are obtained from satellites located to either side of the antenna's main beam (Figure 7–26). If used on a large antenna with suitable f/D characteristics, a multiple feedhorn array will capture sufficient signal even though the secondary focal points are of lesser intensity.

Dual-Band Retrofits

Just about any C-band satellite dish also can be used to receive Ku-band satellite signals. Solid metal dishes tend to perform best, but even perforated or mesh antennas will work—especially if the holes in their surfaces are ex-

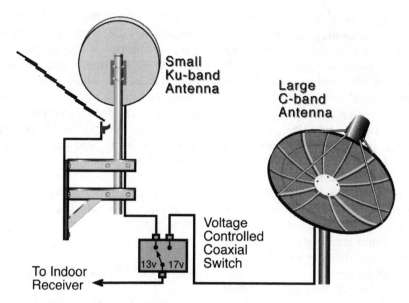

Figure 7–25 A two-dish receiving system using an electronic switch.

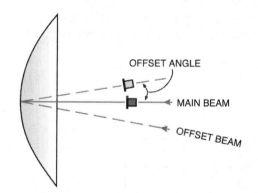

Figure 7–26 The prime focus parabolic antenna creates multiple focal points that are offset at angles away from the antenna's axis of symmetry.

tremely small. One important consideration is the surface accuracy of the dish. Does the antenna surface look smooth and continuous, with little variation in how the mesh material adheres to the frame? Mesh antennas should not have major bumps or depressions in the mesh material. Variations of as little as a frac-

tion of an inch can affect antenna performance. The wavelength of a Ku-band satellite signal is about one-third the wavelength of a C-band signal; surface inaccuracies that will not have much of an affect on C-band signals will affect the antenna's ability to receive Ku-band signals.

It is a relatively simple affair to replace an older C-band feed horn with a new dual-band feed. The existing C-band LNB can be removed from the older feedhorn and attached to the mating C-band flange on the new dual-band feed. The installer may elect to use a remote electronic switch to connect the outputs of the C- band LNB and Ku-band LNB to a single receiver or IRD. In this case, the IRD must be capable of switching its LNB power supply voltage between 13 and 17 volts DC—the essential requirement for triggering the remote band switch installed in proximity to the antenna.

The use of a dual-band feedhorn may not be the best choice for small-dish installations (2 m or smaller) or where the reception quality from one of the popular C-band satellite TV services is already marginal. Dual-band feeds place both the C- and Ku-band feed openings directly at the focal point of the antenna. This placement of both feed openings at such proximity to each other results in some C-band signal loss. If some of your favorite C-band TV services already display impulse noise or "sparklies" in their pictures, then switching to a dual-band feed will only further degrade your reception of these channels.

The Ku-band retrofit feeds are available that bolt directly onto one side of the C-band feedhorn. In this case, the Ku-band feed opening is offset from the C-band feed horn's waveguide and thus will not affect the reception of the C-band channels significantly.

The Ku-band retrofit feedhorn is offset from the C-band feed horn along the plane of the antenna that is perpendicular to the polar axis of the dish. Signals received by the Ku-band retrofit feed will therefore be several decibels lower than if the Ku-band feed opening was directly in front of the dish (see Figures 7–27 and 7–28). The loss of a few decibels may cause little, if any, perceptible change in your reception of higher-powered Ku-band satellite TV services. If you are using a C-band dish that is 3 m in diameter to receive a Ku-band satellite TV signal that normally requires a 60- or 90-cm dish, the system designer can afford to sacrifice a few decibels to maintain optimum C-band reception. Whenever the system switches from C-band to Ku-band transponders on the same satellite, the antenna must move to receive the secondary focal point over which the Ku-band feedhorn is located.

KEY TECHNICAL TERMS

The following key technical terms were presented in this chapter. If you do not know the meaning of any term presented below, refer back to the place in this chapter where it was presented or refer to the Glossary before performing the quick check exercises that follow.

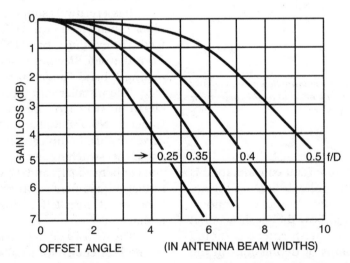

Figure 7–27 *The intensity of the offset beams varies as a function of both the number of wavelengths from the main beam and the focal length to antenna diameter (f/D) ratio of the dish. (Courtesy of Taylor Howard.)*

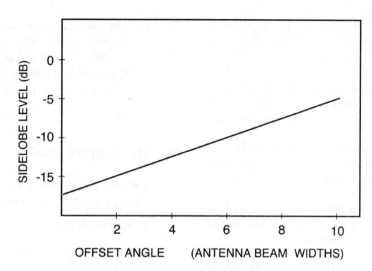

Figure 7–28 *This chart shows the sidelobe levels as a function of offset beam angle and the number of antenna beam widths from main antenna beam to any offset beam. (Courtesy of Taylor Howard.)*

Actuator

Arc zenith

Attenuation

Azimuth

Coaxial cable

Declination

Elevation

Hour angle

Inclinometer

Latitude

Longitude

Magnetic correction factor

Multiple feedhorn array

SCART connector

QUICK CHECK EXERCISES

Check your comprehension of the contents of this chapter by answering the following questions and comparing your answers with the self-study examination key that appears in the Appendix.

Part I: True or False

____ 1. The inclination angle of the feedhorn opening and the inclination angle of the rim of the dish should be the same.

____ 2. The distance from the rim to the dish to the feedhorn's waveguide opening from various points along the rim should be equal.

____ 3. The angle of the modified polar mount's polar axis and the rim of the dish should be the same, regardless of site location.

____ 4. The focal length is shorter for a so-called deep dish than it would be for a shallow dish of the same diameter.

____ 5. The adjustment of the feedhorn's scalar ring plate controls the polarization alignment of the receiving system.

6. The adjustment of the feedhorn's skew sets the polarization alignment to the receiving system.

7. The modified polar mount's declination offset angle is a function of the physical latitude of the site location.

8. The mount's polar axis angle is a function of the physical longitude of the site location.

9. An antenna with a diameter of 3 m and a focal length of 120 cm would have an f/D radio of 0.4.

10. Knowledge of the site's magnetic correction factor is essential in determining the actual inclination angle of the dish.

11. The mount's polar axis should be aligned to true north for locations south of the equator or true south for locations north of the equator.

12. The azimuth angle for any satellite visible from the site location can be determined by using a compass and the inclination elevation angle determined by using a level or plumb bob.

13. To prevent moisture from entering the LNB's F connector, electrical tape should be wrapped tightly around the connector and nearby area of coaxial cable.

14. During the initial site survey, all compass readings should be taken out in the open, away from overhead electrical lines and high-power transformers.

15. Site locations on the Earth's equator will not require *any* declination offset angle.

16. Line amplifiers are recommended to boost signal strength over long lengths of cable because they are cheaper than using a lower loss coax, which is very expensive.

Part II: Multiple Choice

17. Antenna performance can be peaked by taking a signal reading with a:

    a. spectrum analyzer

    b. signal turning meter

    c. volt/ohmmeter

    d. oscilloscope

    e. a + b

18. The antenna mount angle that tilts the antenna downward slightly toward the Earth's equator is called the:

    a. inclination angle

    b. elevation angle

    c. polar axis angle

    d. declination angle

    e. hour angle

19. The antenna mount angle that is a function of site latitude is called the:

    a. illumination angle

    b. diurnal angle

    c. offset angle

    d. declination angle

    e. hour angle

20. To determine the correct azimuth bearing when using a compass, the magnetic factor for the site location should be:

a. added to the compass reading

b. subtracted from the compass reading

c. either a or b, if the site is E or W of true N/S line

d. divided into the compass reading

e. multiplied by compass bearing

21. The arc zenith angle at the site location is:

a. the antenna declination offset angle plus the polar axis angle

b. the offset polar axis angle minus the declination angle

c. 90 degrees minus the polar axis angle

d. 90 degrees plus the polar axis angle

e. none of the above

CALCULATION OF AZIMUTH AND ELEVATION ANGLES

The calculation of the site location's elevation and azimuth angles for any satellite in geostationary orbit requires two constants, two variables, an intermediate equation, and two final equations using trigonometric values.

The constants are $E = 3,959$ (Earth mean radius), and $C = 26,200$ (Clarke Orbit radius). The variables are $A =$ antenna site latitude, $S =$ satellite longitude minus antenna site longitude.

Intermediate calculation (distance between satellite and antenna site):

$$D = [[C \cos(S) - E \cos(A)]^2 + [C \sin(S)]^2 + [E \sin(A)]^2]^{1/2}$$

Final calculations of azimuth and elevation values:

Azimuth = arctan[tan(S)/sin(A)] + 180 (for sites in Northern Hemisphere)

Azimuth = arctan[tan(S)/sin(A)] (for sites in Southern Hemisphere)

Elevation = arccos (($D^2 + E^2 - C^2$) / $2DE$) − 90.

The following computer program in BASIC for DOS (with radian conversion in trigonometric functions) also may be used to calculate azimuth and elevation. (Courtesy of Jim Roberts of Gourmet . . . Entertaining.)

```
10    E=3959
20    C=26200
30    INPUT "SITE LATITUDE":A
40    INPUT "SITE LONGITUDE":LS
50    INPUT "SATELLITE LONGITUDE":LB
60    S=LB-LS
70    D=SQR((C*COS(S)-
      E*COS(A))2+(C*SIN(S))2+(E*SIN(A))2)
80    AZ=ATN(TAN(S)/SIN(A))+180
90    EL=ACN(((D*D)+(E*E)-(C*C))/(2*D*E))-90
100   SK = ATN(SIN(S)/TAN(A))
110   PRINT:AZ:EL:SK
```

Chapter 8

The Satellite/Internet Connection

The convergence of satellite and Internet technologies is one of the more exciting aspects of today's digital satellite revolution. Satellites are proving to be highly effective multimedia platforms because of their ability to deliver information at high data rates to virtually any location within a given signal coverage zone.

Multimedia delivery platforms share many of the features that digital DTH bouquets commonly employ. Several services can be multiplexed onto a single carrier, and fixed-length packets can be used to deliver specific information requests. These packets are addressed so that they can be delivered to specific subscriber terminals as well as encrypted to prevent unauthorized reception.

All satellite/Internet delivery systems are based on an Asynchronous Transfer Mode (ATM) architecture that uses a low-speed modem connection (typically less than 56 Kbaud) for subscriber information requests and a high-speed satellite channel to deliver to each subscriber the requested information.

For example, a data transmission speed of 400 Kb/s over the satellites is more than three times faster than ISDN and 14 times faster than a standard 28.8-Kbaud modem.

Subscribers typically use a terrestrial telephone line for their modem connection, while obtaining their high-speed link from a small satellite dish and a PC adapter card (Figure 8–1) that plugs into one of their computer's expansion ports. Satellite bandwidth is not dedicated to any single user, but rather is shared by all system users on an as-needed basis.

THE DIRECPC AND DIRECDUO SYSTEMS

In 1995, Hughes Network Systems (HNS) introduced its DirecPC (http://www.direcpc.com) satellite delivery system, which provides subscribers in North America with high-speed downloads from the Web. In Europe, Hughes Olivetti Telecom (http://www.direcpc.co.uk)

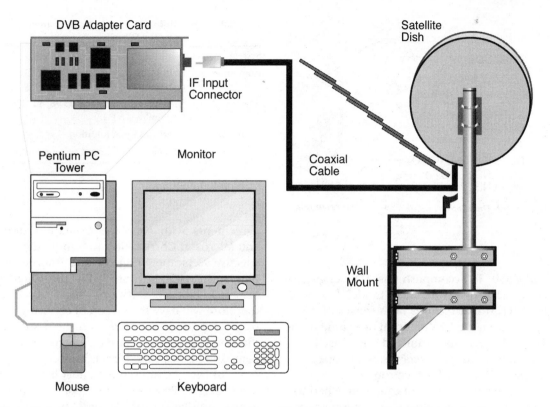

Figure 8–1 *Block diagram of a multimedia work station with satellite delivery of information content.*

uses capacity on the Hot Bird 3 satellite at 13 degrees east longitude to deliver the DirecPC service to subscribers in Europe, Africa, and the Middle East. Hot Bird 3 offers a wide transmission beam that covers Western and Central Europe, Eastern Europe, the Middle East, and North Africa. Within Western Europe, subscribers can receive Internet downloads as well as a variety of TV and radio services on dishes with a diameter of just 60 cm.

In 1997, HNS released its DirecDuo (http://www.direcduo.com) multimedia system, which gives subscribers in the USA access to both high-speed Internet connectivity and the DirecTV bouquet of digital DTH TV services, all from a single fixed satellite dish.

The DirecDuo receiving system is capable of delivering both the 400 Kb/s Internet

service from DirecPC and the 200-plus digital satellite system (DSS) TV and audio channels that DirecTV and USSB offer their digital DTH subscribers. DirecDuo consists of three major components: the DirecPC adapter card that goes inside the subscriber's personal computer, a DSS receiver that connects to the TV set, and an outdoor unit that is installed in the backyard, on the roof, or on the building structure.

The outdoor unit contains a single, 20 × 36-inch elliptical satellite dish equipped with a special low-noise feed (Figure 8–2) and a universal mount. Hughes Network Systems has designed a stationary antenna capable of receiving signals from two separate satellites simultaneously: the DirecPC satellite and the satellite constellation employed by DirecTV

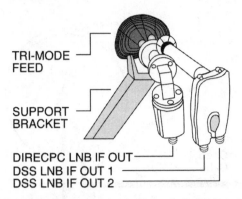

Figure 8–2 DirecDuo tri-mode feed with dual LNB units.

Table 8–1 Minimum System Requirements for DirecPC

System:	Windows 95/98; Windows NT 4.0
Computer:	Pentium Processor based PC, 90 MHz recommended
RAM:	16 Mbytes for Windows 95/98, 32 Mbytes for Windows NT 4.0
Hard disk:	20 Mbytes for application
Modem:	9.6 Kbaud, 28.8 Kbaud recommended

and USSB. To accomplish this, HNS engineers equip each dish with a "tri-mode" low-noise feed (LNF) that receives both senses of polarization from the DSS platform operating in the 12.2–12.7 GHz spectrum and one sense of polarization from the DirecPC satellite operating in the 11.7–12.2 GHz spectrum.

The antenna mount can be attached to a roof, exterior wall or deck, or any other location with an unobstructed line of sight to the south, pointing to the DirecPC satellite. Basic packages include a ground-mount installation with 50 feet of coaxial cable, standard roof or wall-mount installation with 100 feet of coaxial cable, and premium roof, or a wall mount with 100 feet of cable. Special custom antenna installations are available by quote where zoning regulations or other reasons require unusual placement and/or custom bracketing.

The auxiliary hardware and software required to operate DirecPC include an IBM-compatible PC with Pentium processor, Microsoft Windows 95/98 or Windows NT 4.0, at least 16 MB of RAM, 20 MB of free hard disk space, a 9600 modem or better, and an Internet service provider (Table 8–1). DirecPC does not support other computer platforms such as Macintosh or Unix.

An Internet browsing software package also is required. HNS has concluded licensing agreements with Netscape Communications and Microsoft Corporation to bundle their respective Netscape Navigator and Internet Explorer browser software with the DirecPC Personal Edition software. The subscriber therefore will have the option of choosing one of these leading browsers with the assurance that these two browsers have been tested for compatibility with the DirecPC system.

All DirecPC subscribers must have an Internet Service Provider (ISP), a company or organization that provides a local connection to the Internet, either through terrestrial land lines or through a digital cable connection. The ISP typically charges a flat rate for an unlimited number of hours or for a set number of hours each month. DirecPC subscribers either can retain their current Internet Service Provider (ISP) and receive a separate bill from DirecPC, or receive both their ISP and DirecPC services from DirecPC, paying for both on one integrated bill. In addition, bundling the ISP service allows customers to simultaneously configure their DirecPC and ISP services with a click of a button in the user installation menu.

HNS advises subscribers that the components—especially the outdoor unit—should be installed by an installer or technician who is experienced with such similar tasks as installing a satellite antenna or standard TV aerial. This represents an excellent opportunity for satellite professionals to expand their service lines.

DIRECPC SERVICES

HNS offers several different services to its DirecPC and DirecDuo subscribers. Turbo Internet is a two-way interactive service where subscribers use either the Navigator or Explorer web browsers to make their information requests over a modem connected to an ordinary telephone line or ISDN line. DirecPC receives the information request at its Network Operations Center, obtains the data from its IP data gateway to the web, and then relays the data over the satellite link to the subscriber. Called the "pull mode" of data delivery (Figure 8–3), this type of system is typically used to provide subscribers with access to web sites, on-line newspapers, and software downloads. Turbo Internet provides the high-speed Internet connection and supports other Internet capabilities, including Gopher, file-transfer protocol (FTP), e-mail, and Usenet.

HNS also broadcasts multimedia content directly to its subscribers in what is called the "push mode" (Figure 8–4). HNS is responsible for the selection of multimedia content; therefore, no return link is required.

With Turbo Webcast, DirecPC subscribers can choose from a list of the most popular sites on the Internet and have those sites delivered automatically to their hard drives over high-speed satellite links. Because the information is delivered straight to their PC hard drives, access to those sites is instantaneous. Moreover, the delivery process only employs the DirecPC satellite dish, so the subscriber's phone lines remain free for household or

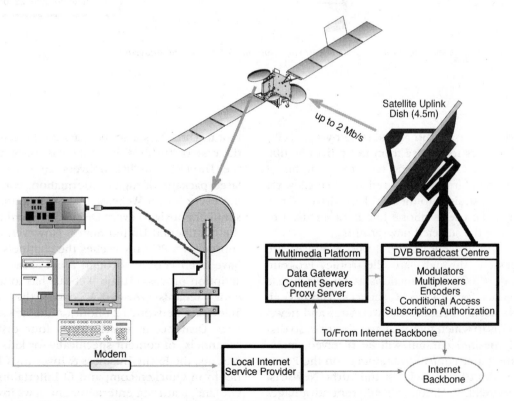

Figure 8–3 High-speed multimedia system featuring "pull mode" delivery of datagrams.

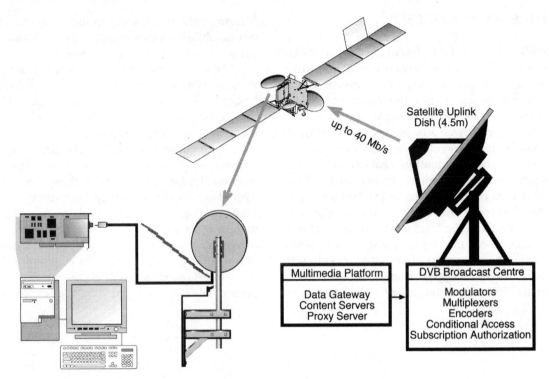

Figure 8–4 *High-speed multimedia system featuring "push mode" delivery of datagrams.*

business use. Turbo Newscast allows DirecPC subscribers to choose from more than 30,000 Usenet newsgroups and have the contents of those newsgroups delivered automatically, via satellite, straight to their PC hard drives. Once again, the user's phone line is never used to retrieve the selected newsgroups.

Turbo Webcast and Turbo Newscast are accessed through an Electronic Program Guide (EPG) that functions as an easy-to-use interface for all DirecPC services. The EPG allows users to choose the web sites and newsgroups to which they wish to subscribe, so that only desired content will be received, maintaining user control over storage on the PC.

Both Turbo Webcast and Turbo Newscast use satellite technology's inherent advantages in broadband bandwidth and point-to-multipoint information delivery to distribute content to DirecPC subscribers across the USA. In the case of Turbo Webcast, at least once a day the DirecPC satellite delivers an extremely large package of digital information, containing every Turbo Webcast "channel" available, simultaneously to every DirecPC subscriber in the continental United States. However, each subscriber's PC only caches the channels that have been chosen through the Webcast subscription process. Turbo Webcast has four anchor tenants: ABCNEWS.com, the ABC Television network's 24-hour online news service; Disney.com, which offers four distinct "channels" of content specifically for kids and parents; the Excite Business & Investing Channel from Quicken.com; and E! Entertainment On-Line, featuring entertainment news involving celebrities, movies, and other TV viewing information.

Within Turbo Webcast, each delivered channel contains a substantial portion (typically 10–30 Mbytes) of the contents of the original Web site, all of which is cached on the user's PC. If the user clicks on a link within a Webcast site that is not cached, Turbo Internet instantly launches into operation, creating a seamless high-speed experience between cached and noncached content.

In contrast, newsgroups in the Turbo Newscast service are continuously delivered as soon as the DirecPC Network Operations Center retrieves them from the Internet. Subscribers can select how much of their PC hard drive capacity they wish their chosen newsgroups to occupy, as well as just when that newsgroup content should expire and be replaced. This is accomplished through the use of the configuration utility that is part of DirecPC's EPG.

DIRECPC SYSTEM INSTALLATION

The DirecDuo dish and the DSS receiver are installed in much the same manner as most digital DTH systems with one exception. Most offset-fed antennas sold for digital DTH reception are installed with their major axis perpendicular to the site location's vertical plane and their minor axis in parallel with the horizontal plane. The 20 × 36-inch elliptical dish designed by HNS, however, must be installed with the major axis in the site location's horizontal plane. The DirecDuo elliptical antenna has the property of exhibiting lower sidelobe response when the major axis lies along the horizontal plane. (To review the antenna and IRD installation procedures for small-aperture antenna systems, see Chapter 7.)

The DirecDuo antenna is a shallow dish that has the property of creating multiple focal points along the antenna axis that is in parallel with the geostationary arc. Because of this property, the "tri-mode" feedhorn receives signals from the satellite that is adjacent to the central satellite at which the antenna's main beam points.

If the subscriber has both the DirecPC adapter and the DSS receiver, he or she can aim the antenna using either device. This is because the DirecPC/DSS LNB units that are part of the antenna feed assembly have been designed so that if the installer aims the antenna successfully at either satellite platform, the antenna is automatically aimed at the other.

SOFTWARE INSTALLATION PREPARATIONS

The DirecPC Personal Edition software installation will require the original Windows 95/98 CD-ROM or Windows 95/98 diskettes to be available during this procedure. Some computers do not come with a separate Windows CD-ROM or set of floppy diskettes. In this case, the installer must know where the Windows files have been installed on the computer's hard disk drive.

To find out where the files are located, run Windows on your computer, click on the "Start" button, then select the "Find" and then the "Files or Folders" commands. Under the "Name & Location" tab, enter the text "*.CAB" in the white box entitled "Named:" and then press the "Find Now" button. A list of all the files with the .CAB extension name will be displayed in the scroll-down window below. The required files will be named something like "Windows95_01.CAB, Windows95_02.CAB," etc. Make a note of the directory path shown for these files (for example, C:\Windows\Options\Cabs\).

ADAPTER CARD AND SOFTWARE INSTALLATION

The installation of the DirecPC adapter card (see Figure 8–5) and software should be relatively straightforward for those individuals who

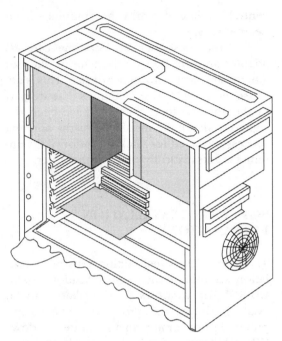

Figure 8–5 Internal view of a pentium-based PC with tower configuration showing PCI and ISA expansion card slots with modem card already installed.

have handled similar procedures for other peripheral computer devices and software. The installer must shut down Windows and then switch off the personal computer, as well as any other peripheral devices that are connected to the system. Before taking the adapter card out of its protective wrapper, discharge any static electricity by touching a metal surface on the computer. Unplug the computer's power cord from the AC wall outlet and then remove the cover from the personal computer.

The DirecPC adapter card is much like any other circuit board or device that can be plugged into a personal computer to add a new capability. For example, the installation of an Iomega Zip 100 or Jaz drive requires the installation of a SCSI card in one of the computer's expansion slots in order for the new

drive to communicate with the computer platform and its software. The Windows "plug-and-play" environment also features automatic detection of newly installed peripheral devices as well as an "installation wizard" that streamlines the process.

Personal computers contain a local bus—the computer's data path between devices—that connects the expansion slots and their cards to the motherboard's central processing unit (CPU). This permits the exchange of data at very fast rates. Two types of expansion slots are available: PCI (for Peripheral Computer Interconnect) and ISA (Industry Standard Architecture). Intel developed the PCI (Peripheral Component Interconnect) local bus for attaching high-speed peripheral devices to the computer system. PCI features a 133 Mbyte/s data transfer rate. The ISA Bus architecture developed for IBM PC and compatibles features a 16-bit data path, up to 3 Mbyte/s data transfer rate, and up to 10 Mbyte/s data transfer rate with bus mastering.

With the cover removed, locate the computer's local bus with its expansion slots. The DirecPC adapter must be installed in one of the computer's empty 32-bit PCI expansion slots. These slots are much shorter than the ISA expansion slots. Moreover, there are three sockets in the 64-bit PCI slot, whereas the ISA slot has only two sockets. Remove the metal cover on the back of the computer for the corresponding PCI slot and slide the mating end of the adapter card into the slot until the card snaps into place. Use the metal screw that was removed from the slot's metal cover to lock the card to the frame of the computer, replace the cover, and insert the computer's power plug back into the AC wall receptacle.

The computer's BIOS (basic input/output system) automatically will configure all the plug-and-play devices, including any PCI bus adapters, each time the machine is restarted. The BIOS is the basic control center for the keyboard, monitor, disk drives and other parts of the computer. Under normal circumstances, the operator never has to do anything to the BIOS unless upgrading the microprocessor or memory, or when updating a really old computer to run newer hardware. If required, the computer will change the BIOS device settings automatically whenever a new plug-and-play device is added to the computer.

Turn the PC on. The Windows operating system will automatically detect that there is a new PCI network adapter inside the computer.

Insert the DirecPC CD-ROM and then click the "Next" button on the "New Hardware Found" dialogue box. The installer may receive a request to install the Windows CD-ROM. If there is no CD-ROM for the machine, the installer will need to type in the path to the *.CAB files on the computer and click "OK." A new dialogue box will then ask for the location of the file "bicndis.sys." If not already present, the DirecPC CD-ROM disk should be inserted into the computer's CD-ROM drive. Click the "OK" button. Windows 95 will conclude the installation process by loading the required files from the DirecPC CD-ROM to the computer's hard disk drive.

Before you install the DirecPC software, exit from all open applications. Restart the computer and insert the DirecPC CD-ROM. The DirecPC welcome screen will appear. Click on the "Next" button and then follow the instructions on the screen to finish the software installation.

The DirecPC software features an "Antenna Pointing" screen that provides site-specific elevation, magnetic azimuth, and polarization values for the dish. These values will be used to complete the antenna alignment. Click on the "Finish" button, and then click "Yes" to restart the personal computer.

Align the antenna using the data provided, and then configure the DirecPC software according to the on-screen instructions. The steps will include the on-line registration of the DirecPC user and the initial setup of the Turbo Webcast service. Once these steps have been completed, the subscriber can begin using the DirecPC Electronic Program Guide (EPG) to access the Turbo Internet, Turbo Webcast, and Turbo Newscast services.

DIRECPC GLOBAL DIGITAL PACKAGE DELIVERY

Satellite delivery of Internet services to businesses around the world also is coming our way. A firm with offices in San Francisco, Mexico City, Edinburgh, Tokyo, and New Delhi will soon be able to distribute a large multimedia report containing graphics, video, and animation to each of its remote offices within hours. DirecPC has launched of a new service called Global Digital Package Delivery that will enable users to send any type of data file, from heavy graphics and text to video, from one site to any number of sites, anywhere in the world, that are equipped with DirecPC satellite receiving systems.

HNS has licensed operators to provide service in Canada, Mexico, Western and Eastern Europe, North Africa and the Middle East, India, Korea, Taiwan, Japan, and numerous other Pacific Rim nations. The DirecPC global rollout was to have been complete by the end of 1998.

The heart of every DirecPC operation, no matter where it is located, is the Network Operations Center (NOC). From the NOC, all direct communications with the satellite take place. To make Global Digital Package Delivery a reality, each existing and future NOC will be operationally interconnected—that is, they will share common billing and delivery prioritization standards.

Through the DirecPC hardware platform, businesses will receive three primary services. Turbo Internet provides Internet access at up to 400 Kb/s. Package Delivery allows point-to-multipoint distribution of data at up to 3Mb/s. DirecPC Multimedia permits point-to-multipoint distribution of full-screen, MPEG-1 quality video straight to the computer desktop.

DVB-COMPLIANT MULTIMEDIA SYSTEMS

Both the EUTELSAT (http://www.eutelsat.org) and ASTRA (http://www.astra.lu) satellite operators intend to offer new multimedia satellite platforms for Europe, North Africa, and the Middle East in the near future. These new platforms will be based on the open architectures of the DVB and MPEG-2 digital standards and will feature high-speed Internet access at data rates of up to 2 Mb/s per individual user (see Figure 8–3) or 40 Mb/s for delivery of data broadcast services (see Figure 8–4).

Europe's DVB Group has selected the MPEG-2 DSM-CC (Digital Storage Media— Command and Control) specification to serve as the heart of the DVB Data Broadcasting Standard in conjunction with DVB-SI (Service Information). This DVB data broadcasting specification is based on a series of four Profiles, with each Profile corresponding to a particular application area for DVB-compliant broadcast networks.

The Data Piping Profile is for the simple, end-to-end delivery of anonymous, nonsynchronous digital bitstreams. The Data Streaming Profile is for the end-to-end transport of asynchronous, synchronous, or synchronized bitstreams. The Multiprotocol Encapsulation Profile permits the use of the DVB transport mechanism for different communications services that require the transmission of communication protocol datagrams. The Data Carousel Profile supports the periodic transmission of comprehensive data modules. In addition to these four Profiles, an object carousel specification has been added in order to support data broadcast services that require the broadcasting of objects as defined in the DVB Network Independent Protocols specification. The equipment required consists of a satellite dish and a DVB/MPEG-2 adapter card for the computer. Antennas that already receive digital TV services from the EUTELSAT and ASTRA satellite constellations also can be used to receive high-speed multimedia services. The return link consists of an ordinary modem connected to a telephone line or an ISDN line.

Each multimedia service provider has a Network Control Center (NCC) that acts as a DVB broadcast station. The NCC consists of modulators, multiplexers, encoders, and a system controller as well as a conditional access and subscriber management system. The DVB platform connects to a multimedia platform consisting of satellite servers/routers and a data gateway using TCP/IP (Transmission Control Protocol/Internet Protocol). TCP/IP is a suite of protocols that allows communications between groups of dissimilar computer systems from a variety of vendors. The IP part of TCP/IP is the protocol used to route a data packet, called a "datagram," from its source to its destination over the Internet. The platform performs the elementary tasks of a server and distributes IP datagrams as an MPEG-2 data structure in the DVB-compliant MPEG-2 digital bitstream.

Digital bouquet operators soon will be able to take advantage of the fact that digital data can be multiplexed with their digital DTH TV services. Service providers therefore will be able to obtain new revenue streams by offering Internet content as well as new data broadcast services to their subscribers.

INTERNET GROWTH IN ASIA

The satellite delivery of Internet content represents one of the fastest growing uses of satellite capacity in the Asia/Pacific region. At the present time, these Internet transmissions are largely confined to international satellite systems such as INTELSAT and PanAmSat, which offer the regional or national Internet Service Provider (ISP) a "backbone" service that can provide high-speed interconnectivity between Asia and North America. These backbone services are asymmetric, that is, configured to deliver digital information at a higher rate in one direction (North America to Asia) over the other (Asia to North America). This is because more than 70 percent of the web sites to be found on the Internet today are located on computer servers based in the United States.

Within the past year, however, several Asian companies have begun to offer their own satellite connections to the World Wide Web. DirecPC Japan, for example, uses the Superbird C satellite to deliver Internet content to Japanese businesses. Users even have the ability to select and download video programs at speeds of up to 3 Megabits per second.

Thailand-based CS Communications, a subsidiary of the Shinawatra Group, has launched a new service for rural businesses and organizations that provides a direct Internet connection via Ku-band capacity on the Thaicom satellite system. Called Thaicom Direct, the new service provides a package solution that includes transmission connections, a satellite dish, and Internet software tools. The new service, for example, will permit the establishment of Internet cafes in tourist spots that previously have been unserved.

Regional service provider Zaknet recently began offering an Internet multicasting service on the AsiaSat 2 satellite that allows its subscribers to download 300 of the Internet's most popular web sites, as well as regional newspapers, stock quotations, and financial reports. No return telephone connection is required because Zaknet preselects the Internet content that it broadcasts. The Zaknet service also includes the video streaming of live news coverage from TV programmers such as CNN and Bloomberg. All web sites contained in the Zaknet multicasting service are continuously updated to ensure that users get the most upto-date information available. Zaknet currently markets its multicasting service in individual Asian countries through a network of value added resellers.

CABLE'S INTERNET OPTIONS

Cable and SMATV system operators now have the option of offering high-speed Internet services to their subscribers as a way to generate additional revenues and attract new customers. In 1998, Scientific-Atlanta introduced Worldgate, a new multimedia delivery system that includes a set-top box for cable and SMATV applications that are equipped with software that allows Internet web pages to be displayed on a standard TV screen.

The Worldgate system centralizes all computer processing at the cable or SMATV system's head end. With Worldgate, the existing store-and-forward addressable communications system that the subscriber normally uses to order pay-per-view movies and events has been altered to operate in real time so that each subscriber can select web content using the same remote control used to watch TV programming. Subscribers use an on-screen keyboard and pointing device to enter web addresses and create e-mail. A low-cost wireless keyboard is also available as an option for subscribers who expect to make extensive use of the Worldgate system's e-mail and surfing capabilities.

With Worldgate, cable subscribers can log onto the Internet almost instantaneously and download web pages at approximately three times the speed of the fastest telephone modem when equipped with an analog-based cable set-top box, or 1,000 times faster with a digital set-top box. Since the cable provides both the forward and return links, no connection to a telephone line—with the resulting additional costs and speed limitations—or an outside Internet Service Provider (ISP) are required. Best of all, the Worldgate software allows cable subscribers to instantaneously go from any given cable TV program service to its associated web site. The Worldgate Internet Client Service installed at the cable system's head end matches the channel, time, and program data to the corresponding web site, eliminating the need for the individual subscriber to enter the TV service's web site address or URL.

The Worldgate system consists of a head-end server that is designed to operate in either one-way or two-way plants and Scientific-Atlanta's standard analog or digital set-top converters. The head-end server provides the link between the cable system's subscribers and the Internet. The cable TV system operator may elect to include a satellite link for high-speed access to the Internet or a high-speed terrestrial land-line or fiber-optic connection. Popular web sites can be stored or "cached" on the cable TV system's head-end server for the fastest possible subscriber connections and downloads.

KEY TECHNICAL TERMS

The following key technical terms were presented in this chapter. If you do not know the meaning of any term presented below, refer back to the place in the chapter where it was presented or refer to the Glossary before trying the subsequent quick check exercises.

Adapter card

Asynchronous transfer mode (ATM)

BIOS

Internet Service Provider

Industry Standard Architecture (ISA)

ISDN

Modem

MPEG-2 DSM-CC

Peripheral Computer Interconnect (PCI)

Pull mode data delivery

Push mode data delivery

Transmission Control Protocol/Internet Protocol (TCP/IP)

QUICK CHECK EXERCISES

Check your comprehension of the contents of this chapter by answering the following questions and comparing your answers to the self-study examination key that appears in the Appendix.

Part I: Matching Questions

Put the appropriate letter designation—a, b, c, d, etc.—for each term in the blank before the matching description.

a. TCP/IP

b. PCI expansion slot

c. ISA expansion slot

d. modem

e. pull mode

f. push mode

g. ISP

h. Internet adapter card

i. ATM

j. datagram

k. MPEG-2 DSM-CC

1. The _____ architecture uses a low-speed _____ connection to initiate subscriber requests and a high-speed satellite channel to deliver the requested information to the subscriber.

2. The pull mode of delivery requires that each subscriber have a local _____ connection for the initiation of all subscriber information requests.

3. _____ is a suite of protocols that allows communication between groups of dissimilar computer platforms.

4. The _____ is a data packet that carries information from its source to the destination over the Internet.

5. The _____ specification serves as the heart of the DVB data broadcasting standard.

6. A _____ is the communications device that allows personal computers to connect to the Internet over traditional telephone lines.

7. The one-way satellite broadcast of multimedia content to subscribers is known as _____.

8. The two-way interactive broadcast of multimedia content to subscribers is known as _____.

INTERNET HYPERLINK REFERENCES

Technical notes on the DirecPC satellite delivery system.
(In the USA):
http://www.direcpc.com/about/nav_howwork.html

(In Western Europe):
http://www.direcpc.co.uk/

(In Japan):
http://www.direcpc.superbird.co.jp/

Technical notes on the DirecDuo satellite delivery system.
http://www.direcduo.com/about/how.html

Internet access via the EUTELSAT satellite system (388 K PDF file).
http://www.eutelsat.org/emp.pdf

Chapter 9

Digital SMATV System Overview

SMATV (Satellite Master Antenna Television) systems are community TV receiving systems that allow multiple users to share the same satellite and terrestrial TV program resources. SMATV systems feature the communal use of a dish and private cable distribution network, which significantly reduces the equipment cost per household.

SMATV systems are the preferred mode for cost-effective coverage in a multiple dwelling unit (MDU) environment. The digital SMATV system's communal satellite antenna receives digital DTH program bouquets from one or more satellites, converts the signals into a cable-compatible delivery format, and then sends the converted signals down a coaxial cable to each individual dwelling. A terrestrial TV aerial also may be included in the system to receive the available off-air TV channels. The terrestrial and satellite TV signals are then combined for distribution to households located in one or more adjacent buildings.

SMATV offers several distinct advantages over dedicated DTH receiving systems. The communal use of a single dedicated antenna eliminates the unsightly spectacle of numerous dishes sprouting up on a single apartment building or condominium. Moreover, a larger antenna can be used than what typically is employed for home satellite TV use, which will provide a higher signal margin. Best of all, the SMATV system can combine satellite and terrestrial TV channels efficiently so that all residents in a multiple unit dwelling can access the available programming by means of a simple cable connection.

SMATV SYSTEM COMPONENTS

Like the individual home satellite receiving system, the SMATV system uses a parabolic antenna to receive multiple TV and sound services from a single satellite or constellation of colocated satellites. The major difference, however, is that SMATV systems commonly employ a larger antenna aperture than what DTH systems normally require (Figure 9–1). The purpose of boosting antenna size is to generate a very good signal at the system "head end" to reduce any degradation that might be introduced by subsequent signal processors as well as by the cable distribution system itself.

The SMATV "head end" is the central processing center for all of the signals being received by the satellite antenna as well as one or more terrestrial TV antennas. Depending on the precise configuration of the system, the head end either may convert the incoming satellite and terrestrial channels to an alternate modulation format for subsequent cable distribution or act as a primary distribution amplifier that is totally transparent to the modulation format of the incoming satellite signal.

In either case, the broadband multiplex leaving the head end carries all of the desired TV and sound channels. The coaxial cable distribution system carrying this multiplex must be divided, or "split," several times before it reaches every TV set in the network. To make up for this signal loss, line amplifiers are added at strategic locations along the cable line to boost the signal level back to acceptable levels.

A set-top box will be required at each receiving location within the digital SMATV system. The purpose of the set-top box is to convert the digital multiplex sent over each cable TV channel to individual TV and/or sound channels.

DVB-COMPLIANT SMATV SYSTEMS

The Digital Video Broadcasting (DVB) Group has established several specifications governing the use of the MPEG-2 digital compression standard for broadcasting purposes. The DVB specifications provide for the cross-platform portability of digital video, audio, and data signals so that any DVB-compliant bitstream can be transported from one operating environment to any other without requiring any baseband interfacing. (By baseband I mean a frequency band containing information, either prior to the modulation of the information onto a radio frequency carrier or following demodulation at the receive end.)

DVB-C is a cable broadcasting system for television, sound, and data services using standard cable TV distribution frequencies and bandwidths. DVB-C uses quadrature amplitude modulation (QAM), a form of amplitude-shift keying where the amplitude and the phase of a series of baseband pulses are modulated to represent the message (Figure 9–2). QAM is used for cable distribution systems because it is more spectrum-efficient than QPSK in bandwidth-constrained cable and SMATV environments. A single 8-MHz-wide European cable TV channel, for example, can accommodate a payload capacity of 38.5 Mbit/s if 64-QAM is used as the modulation scheme. The "64" in 64-QAM refers to the number of discrete signal-state values of vector magnitude that the QAM signal supports. Other levels of QAM, such as 16-QAM, 32-QAM, and 128-QAM, also may be employed.

DVB-CS is the DVB specification adapted from DVB-C (cable TV) and DVB-S (satellite) that is applicable to SMATV installations. The DVB-CS standard codifies several different methods for adapting digital signals for

Satellite EIRP:		52	50	48	44	dBW
No. of IRDs in SMATV system:						
1 - 4		60	75	90	120	
5 - 16		75	90	120	150	
17 - 50		90	120	150	180	
Astra recommended size of satellite dish in cm						

Figure 9–1 Ku-band SMATV system recommended antenna sizes.

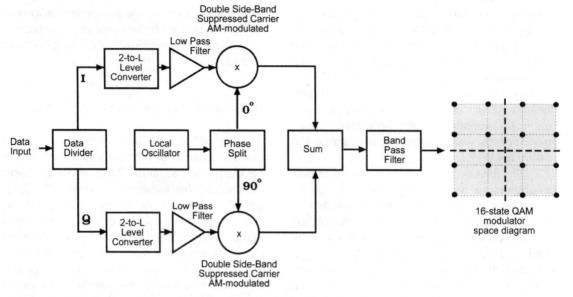

Figure 9–2 *Classical quadrature amplitude modulator (QAM) block diagram.*

distribution through SMATV systems. Each of these methods is tailored to conform to the technical characteristics of bandwidth-limited SMATV channels.

SMATV-DTM

The SMATV-DTM configuration uses QAM rather than QPSK for its modulation scheme. SMATV-DTM features transparent "digital transmodulation" without baseband interfacing for the conversion of QPSK-modulated digital satellite signals to equivalent QAM-modulated signals. The unit at the SMATV system's head end that performs this function is called the transparent digital transmodulator (TDT).

The TDT converts the QPSK-modulated digital bitstream arriving from one satellite transponder into an equivalent QAM signal that can be carried by a cable TV channel that is 6 MHz (North America) or 8 MHz (Europe) wide. All residences connected to the SMATV system must be equipped with a digital set-top

box that is capable of processing these QAM-modulated signals.

Whereas analog-based SMATV systems must use a separate IRD and associated RF modulator for each incoming satellite TV service, the digital SMATV head end merely needs one separate transmodulator to receive each QPSK-modulated satellite transponder, regardless of the number of video, audio, and data services that the satellite transponder carries.

A single 27-MHz-wide satellite transponder may carry a digital multiplex containing six or more TV program services with associated sound channels, along with auxiliary audio and data services. A digital SMATV head end with just 10 transmodulators therefore has the ability to deliver more than 60 digital TV services plus numerous audio and data services over 10 cable TV channels.

The use of TDT units is the most cost-effective option for buildings and condominiums containing more than 80 households, or housing estates with 80 or more individual or semidetached houses. The high cost of implementing a system such as the one just described makes it an impractical solution for smaller SMATV system operators to implement. Two alternative configurations, known as SMATV-IF and SMATV-S, are available for the design of smaller SMATV systems (Figure 9–3).

New digital TV standards introduced in the United States and Europe permit terrestrial TV stations to broadcast as many as five STV (for standard TV) or two HDTV (high-definition TV) services over a single terrestrial TV channel. The DVB-compliant digital SMATV head end also can receive QAM-modulated off-air signals and send them to each dwelling without making any changes to the modulation format. The signals, which are merely converted to a standard cable TV channel frequency, can be decoded by a QAM-compatible set-top box or digital TV set at each residential location in the system. In the United States and elsewhere, a modulation system using vestigal sideband (8-VSB) has been adopted for the terrestrial transmission of digital TV signals. In this case, the 8-VSB signal would have to be transconverted to QAM at the head end before it could be decoded by a QAM-compatible set-top box at each residential location.

SMATV-IF

Both the SMATV-IF and SMATV-S configurations are based on the use of QPSK modulation, where digital satellite signals are received from the satellite and then frequency converted to an IF frequency band that is appropriate for SMATV distribution. As in the SMATV-DTM system previously described, the SMATV-IF and SMATV-S head ends do not make any changes to the baseband characteristics of the digital satellite signal. All of the set-top boxes in the system connect to a single antenna at the head end that is pointed at the satellite constellation of choice. Terrestrial TV signals also can be multiplexed onto the same

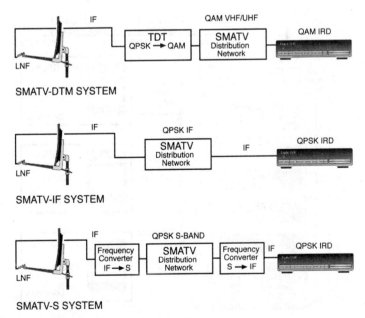

Figure 9–3 SMATV-DTM, SMATV-IF, and SMATV-S signal distribution methods.

coaxial distribution line using the VHF, mid-band and UHF TV frequency channels located below 950 MHz—the point at which the satellite signal distribution begins.

The SMATV-IF specification calls for the distribution of QPSK-modulated satellite signals at the standard block IF output (950–2050 MHz) of the SMATV antenna's Low-Noise Feed (LNF). The main advantage for the system operator is that many of the components normally found in a traditional SMATV head end, such as satellite receivers, decoders, and associated RF modulators, are no longer required. Instead, a processed IF distribution system can be implemented at the SMATV head end that frequency converts all transponders transmitted by a satellite or group of colocated satellites and then distributes them over a single coaxial cable (Figure 9–4).

This "processed intermediate frequency distribution" method is an ideal option for new buildings or condominiums with fewer than 80 dwellings. The so-called "stacked" LNF produces the required wide-band IF output, which contains signals using both senses of orthogonal polarization that the desired satellite employs. The lower range from 950 to 1,450 MHz contains all signals of one polarization, while the upper range from 1550 to 2,050 MHz contains all signals of the opposite sense of polarization. Each residence connected to the digital SMATV system must be equipped with a digital IRD that is capable of receiving and processing the entire range of signals contained within the 950–2,050 MHz IF output band.

Each subscriber also will need to purchase one or more digital satellite IRDs, depending on the number of TV sets in the home to be covered (Figure 9–5). This raises the initial installation cost per customer for signing up. The end result, however, is that each dwelling has direct access to high-quality digital TV and audio services.

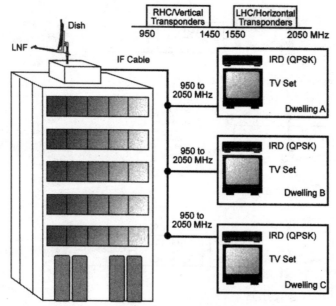

Figure 9–4 *SMATV-IF system block diagram showing dual polarization, single IF output from the head end.*

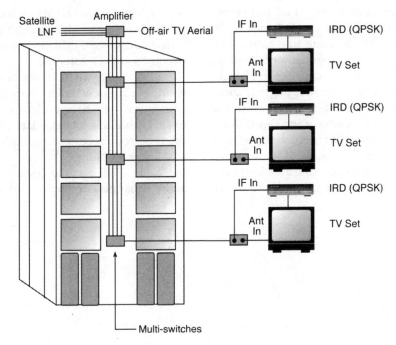

Figure 9–5 *SMATV Multiswitch IF system block diagram.*

SMATV-S

The primary drawback to the processed intermediate frequency distribution method is that the wide-band 950–2,050 MHz IF signal requires more expensive line amplifiers, splitters, and other in-line devices than traditional SMATV systems typically employ. Increased cable attenuation losses will come into play, which will require the use of low-loss coaxial trunk lines and perhaps a greater number of line amplifiers, depending on the distances to be covered.

The SMATV-S head end is a cost-effective alternative that converts the digital satellite signal to a much lower IF frequency range (230–470 MHz in Europe) that can be sent through an existing master antenna television (MATV) distribution network that handles lower-frequency terrestrial TV signals. Each residence also must be equipped with a fre-

quency converter that transforms the low-frequency IF back to the standard IF frequency range that the digital satellite IRD commonly employs.

THE MULTISWITCH IF DISTRIBUTION METHOD

Satellite systems—such as the Astra and Eutelsat satellite constellations currently serving Europe—now offer digital DTH services using Ku-band frequencies that may extend all the way from 10.7 to 12.75 GHz. It therefore is not possible to use either the SMATV-IF or SMATV-S methods previously described to send all of the available signals down a single coaxial cable. Instead, the SMATV antenna must be equipped with a special universal LNF that has four IF output ports. Each pair of IF outputs is dedicated to one of the available Ku-band frequency ranges, either 10.7–11.7 GHz (Low Band) or 11.7–12.75 GHz (High Band). Each IF port outputs signals using one of the two available senses of orthogonal polarization. The IF outputs of the LNF connect to a multiswitch for each building floor. A single IF cable connects the multiswitch for each floor to the digital IRD at each residence. (See Figure 9–6.)

In some systems, the digital IRD selects either the low-band or high-band IF output of the universal LNF by generating a 22-kHz tone that is sent up the coaxial cable to the multi-

switch. The digital IRD also selects the desired polarization sense by switching the voltage sent up the coaxial cable to the multiswitch between 13 and 17 volts DC.

The multiswitch IF distribution method is suitable for new buildings or condominiums with fewer than 20 households that have ducting available to accommodate the insertion of the required cables and multiswitches. Terrestrial TV services may also be multiplexed onto the cable. Wall receptacles are available that allow separate connections to each IRD and

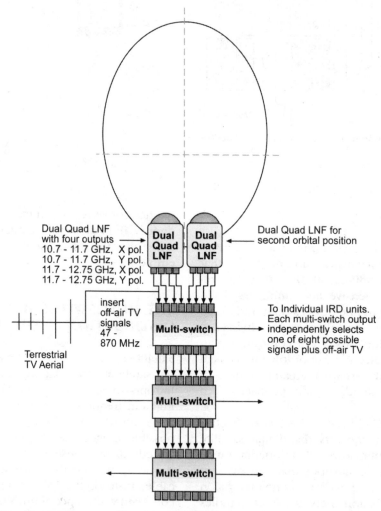

Figure 9–6 SMATV multiswitch IF two satellite system block diagram.

TV set in the system so that each residence has access to all available terrestrial and satellite TV services.

The multiswitch IF distribution method also can be used by those SMATV systems that receive digital signals from multiple satellites located at different orbital assignments. The SMATV system can employ a paraboloid equipped with multiple feedhorns or LNF units so that a single antenna can receive two or more adjacent satellites at the same time. Any paraboloid—including the offset elliptical reflector—that employs a relatively shallow curvature has the ability to generate multiple focal points. Each secondary focal point receives signals arriving from angles that are offset from the reflector's axis of symmetry (Figure 9–7).

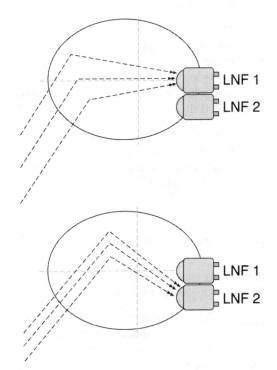

Figure 9–7 SMATV antenna using multi-LNF approach for simultaneous reception of two satellites.

FIBER-OPTIC SMATV SYSTEMS

Fiber-optic solutions for the transmission of digital DTH services in a multiple dwelling unit (MDU) environment also are available. Traditional coaxial cable distribution systems are difficult and expensive to implement in large buildings and garden-style property developments. Fiber-optic distribution systems offer cost-effective distribution to a large number of subscribers from a single antenna.

In this case, the SMATV system employs an L-band fiber-optic transmitter and receiver that has been designed specifically for MDU applications. A single fiber-optic transmitter can send both L-band polarizations to compatible receivers located at secondary distribution points throughout the property. The associated fiber-optic cables may be thousands of feet in length without significant signal attenuation. Typically less than 1 dB of RF insertion loss is added, even at 2,050 MHz. Diplexers also may be used to combine the satellite and off-air signals together in conjunction with the usual network of splitters and/or taps used to connect to individual apartments or condominiums.

KEY TECHNICAL TERMS

The following key technical terms were presented in this chapter. If you do not know the meaning of any term presented below, refer back to the place in the chapter where it was presented or refer to the Glossary before trying the quick check exercises that follow.

Digital transmodulation

DVB-C

DVB-CS

Head end

Multiswitch intermediate frequency distribution

Quadrature amplitude modulation (QAM)

Quadrature phase shift keying (QPSK)

Processed intermediate frequency distribution

Satellite Master Antenna Television (SMATV)

Signal state values

Stacked LNF

Transparent digital transmodulator

QUICK CHECK EXERCISES

Check your comprehension of the contents of this chapter by answering the following questions and comparing your answers to the self-study examination key that appears in the Appendix.

Part I: Matching Questions

Put the appropriate letter designation—a, b, c, d, etc.—for each term in the blank before the matching description.

a. QAM

b. QPSK

c. transmodulation

d. head end

e. baseband

f. processed IF distribution

g. multiswitch IF distribution

h. signal-state values

i. DVB-C

j. DVB-CS

k. stacked LNF

l. TDT

1. The _____ method uses a stacked LNF to produce a wide-band IF signal containing signals of both senses of polarization. This signal is sent to each dwelling's IRD via a single coaxial cable.

2. The SMATV _____ is the central processing center for all signals that are received by the system's terrestrial and satellite antennas.

3. _____ is a form of amplitude shift keying where the amplitude and the phase of a series of baseband pulses are modulated to represent the message.

4. The _____ is the system component that converts the incoming satellite signals from a _____ format to a quadrature amplitude modulation (QAM) signal for distribution to individual residences connected to the SMATV system.

5. The frequency band containing information, either prior to the modulation of the information onto a radio-frequency (RF) carrier or following demodulation of the signal at the IRD, is called the _____.

6. The "32" in 32-QAM refers to the number of discrete _____ of vector magnitude that the QAM-modulated signal supports.

Part II: True or False

Mark each statement below "T" (true) or "F" (false) in the blank provided.

_____ 7. The purpose of the transparent digital transmodulator (TDT) is to convert QAM signals into a QPSK modulation format.

_____ 8. DVB-CS is a set of specifications pertaining to SMATV systems that was established by the Digital Video Broadcasting (DVB) Group.

___ 9. QAM is a more effective modulation scheme than QPSK for the distribution of digital signals in a bandwidth-limited environment such as SMATV, where the bandwidth of a single channel is limited to just a few megahertz.

___ 10. Digital transmodulation is the process whereby signals of opposite polarization from a given satellite are multiplexed onto a single coaxial cable through the use of a transparent digital transmodulator at the SMATV head end.

Part III: Multiple Choice

Circle the letter—a, b, c, d, or e—for the choice that best completes each of the sentences that appear below.

11. The Digital Video Broadcasting (DVB) Group has established the following specifications governing the use of the MPEG-2 digital compression standard for SMATV distribution purposes:

   a. DVB-CS

   b. DVB-S

   c. SMATV-QAM

   d. SMATV-QPSK

   e. a and b

12. An all-digital SMATV head end may contain the following components:

   a. transparent digital transmodulator (TDT)

   b. IRD

   c. Standards converter

   d. RF modulator

   e. a and b

13. The SMATV system normally uses a larger satellite antenna than would be required at for a home satellite TV system because

   a. SMATV systems usually receive their signals from low-power geostationary satellites

   b. the SMATV system must generate a very good signal at the system head end to reduce attenuation and other degradations in the cable distribution system

   c. the larger dish will give the system a greater signal margin to counteract rain fades

   d. a high signal level from the satellites is essential whenever terrestrial and satellite signals are combined on the same cable distribution network

   e. b and c

14. The "16" in 16-QAM refers to

   a. the 16 bits of information normally used in MPEG-2 digital compression block coding systems

   b. the number of variables that are present in the transparent digital transmodulator's digital mixer circuit

   c. the maximum number of megabits per second that the digital encoder can send at any given moment of time

   d. the number of discrete signal-state values of vector magnitude that the QAM signal supports

   e. the amount of bandwidth in megahertz that the QAM signal will occupy on a cable distribution system

15. The transparent digital transmodulator converts the satellite's QPSK-modulated digital signal into an equivalent

a. DVB-compliant signal that contains fewer bits of information than the original signal

b. QAM signal that can be distributed through a 6-MHz-wide (North America) or 8-MHz-wide (Europe) cable TV channel

c. QAM signal that can be retransmitted over another satellite

d. DVB-compliant signal that can be more efficiently distributed over a twisted-pair telephone line

e. all of the above

INTERNET HYPERLINK REFERENCES

EUTELSAT Technical Recommendations for Manufacturers of SMATV Receiving Equipment (Format PDF - 290 Kb).
http://www.eutelsat.org/press/tv_recept1.html

HISPASAT SMATV Reception in Spain/Europe.
http://www.hispasat.com/util/rec_1i.htm

The ASTRA Satellite System: What you need for SMATV reception.
http://www.astra.lu/recept/what/index.htm

Chapter 10

On the Road to HDTV

High-definition television (HDTV) transmissions are capable of presenting approximately twice the number of active lines (1,080) that analog TV pictures currently provide. The video images have a sharpness that approaches the clarity of 35-millimeter film and are presented in a wide-screen, cinemascopic format with an aspect ratio (picture width to picture height) of 16:9 as opposed to the 4:3 aspect ratio used by conventional analog TV transmission systems. HDTV also features the broadcast of TV programs in stereo with surround sound.

During the early 1980s, Japanese broadcaster NHK conducted a series of experimental HDTV broadcasts of its analog-based MUSE HDTV system using Japan's "Yuri" DBS satellite platform. The primary drawback to this early effort to improve upon the four-decade-old PAL, NTSC, and SECAM TV standards, however, was the large amount of bandwidth required to broadcast a single HDTV service. NHK demonstrated the MUSE system to the U.S. Congress and Federal Communications Commission (FCC) during the mid-1980s in an attempt to gather support for the adoption of MUSE technology as a new international TV broadcasting standard. In the end, however, the FCC decided to wait until a more spectrum-efficient technology was available. The development of digital compression technologies during the early 1990s proved to be the key to making HDTV a reality for TV viewers around the world.

HDTV is just one component of new digital TV standards approved within the past two years by the FCC and the International Telecommunication Union (ITU). These new standards promise to eliminate the flaws inherent in the PAL, NTSC, and SECAM TV systems. For example, the new global standard for terrestrial broadcasting will accurately portray all the colors of the original image as well as employ sophisticated digital filtering and forward error correction (FEC) techniques to detect and mask out noise, ghosting, and electrical interference from automobiles and electronic appliances. Video "crawl" and other analog TV picture artifacts will also be a thing of the past.

A NEW GLOBAL DIGITAL STANDARD

On May 30, 1997, the International Telecommunication Union agreed on a new global standard for digital terrestrial television broadcasting (DTTB) that promises to deliver end-to-end digital TV with high-definition quality, and also unify television broadcasting systems worldwide. DTTB represents the construction of a digital architecture that simultaneously can accommodate both high-definition television and conventional standard-definition television services in the terrestrial broadcasting environment, at the same time being interoperable with cable delivery, satellite broadcasting, and recording media.

The ITU also unanimously agreed on the convergence toward a single HDTV production standard based on a High-Definition Common Image Format (HD-CIF) that is characterized by using a single matrix of samples (1,920 pixels by 1,080 lines) irrespective of field and frame rate. This has given equipment manufacturers the go-ahead to start delivering TV sets to anywhere in the world, thus providing economies of scale never available before, as well as worldwide portability for consumers and vendors.

THE HD-CIF FORMAT

The ITU recommendation also unifies two competing standards: the U.S. Advanced Television Standards Committee (ATSC) proposal and the European Digital Video Broadcasting (DVB) proposal. Under the ITU Recommendation, the two systems will form a single compatible system that can be implemented on a global basis within the practical physical limitations of the current terrestrial TV channel assignment environment. Moreover, the new digital system will support multi-program transmissions in existing channels through the use of digital video compression technology.

Analog-based terrestrial TV systems leave adjacent TV channels unoccupied to prevent interference between TV stations operating within the same general broadcast coverage area. It has been determined, however, that the new digital TV services could occupy these unused channels without causing interference to existing analog TV stations. National telecommunications authorities therefore will not need to assign any channel frequencies before introducing DTV services, thereby conserving scarce spectrum resources. Moreover, total use of the frequency spectra assigned for terrestrial TV broadcasting worldwide finally can be utilized in an efficient manner.

Under the ITU plan, existing analog TV transmissions eventually will be phased out (within a ten-year time frame as proposed in the US or within a longer period as envisioned for Europe). As terrestrial TV transmissions change from analog to digital, analog TV sets will be fitted with set-top boxes to enable them to decode and process the new digital TV signals. Chips manufacturers already have announced that they are ready to start mass production of the chips required by the decoders to be integrated in the new TV sets. There currently are 1,288 million TV sets worldwide that eventually will need to be replaced. This is a huge market that represents a golden opportunity for those who work in the consumer electronics industry.

MPEG-2 PROFILES, LEVELS, AND LAYERS

The MPEG-2 compression standard is a key component of the new digital TV standards adopted by the FCC and the ITU. As we previously discovered in Chapter 2, MPEG-2 is actually a family of systems, with each system having an arranged degree of commonality and compatibility.

MPEG-2 supports four different Levels: High, High-1440, Main and Low Level.

The High and High-1440 Levels support high-definition (HDTV) and advanced-definition TV (ADTV) pictures with 1,920 × 1,080 and 960 × 576 sample matrices, respectively. Both of these Levels support two spatial resolution Layers, respectively, called the Enhancement Layer and the Base Layer, which broadcasters can use to deliver standard-definition TV (SDTV) signals, as well as ADTV or HDTV signals simultaneously. This is accomplished by using the low-resolution Base Layer to deliver an SDTV signal with a 4:3 aspect ratio while at the same time using one or more Enhancement Layers to deliver the additional data required to produce higher resolution TV pictures with the wider 16:9 aspect ratio. Together, the enhancement and low-resolution Layers deliver all the information that the HDTV set needs to produce a high-resolution picture. SDTV sets receive the data they require exclusively from the Base Layer, while ignoring the data contained in one or more Enhancement Layers.

All digital bitstreams and set-top boxes are classified according to video frame rate, either 25 or 30 frames per second, depending on the accepted standard in each country of operation. Set-top boxes with dual frame rate capabilities also are possible. Although digital bitstreams are set for one of the two frame rates, an MPEG-2 transport stream may carry video program material that is intended for more than one type of digital TV set or set-top decoder.

MPEG-2 also supports five different Profiles: Simple, Main, SNR Scalable, Spatial Scalable and High. Each profile consists of a collection of compression tools. For example, a Main Profile may use up to 720 pixels per line at Main Level, or up to 1,920 pixels per line at High Level. The new DTV standard adopted by the FCC for use in the United States will operate at Main Profile, High Level (MP@HL).

AMERICA'S GRAND ALLIANCE

In 1983, the Advanced Television Systems Committee (ATSC) was formed to coordinate the technical details of implementing a new Advanced Television (ATV) standard for the United States. In 1987, the Federal Communications Commission (FCC) responded to requests from U.S. broadcasters by initiating an ATV rulemaking and establishing an FCC Advisory Committee on Advanced Television Service (ACATS) for the purpose of recommending a new TV broadcast standard for the United States.

The FCC initially defined ATV as any system that results in improved television audio and video quality. Between 1987 and 1995, hundreds of companies and organizations worked together within the numerous subcommittees, working parties, advisory groups and special panels of ACATS to develop a competitive process by which TV system proponents were required to build prototype hardware that would then be thoroughly tested. Along the way, the FCC made several key spectrum decisions concerning the introduction of a new U.S. television standard. In 1990, for example, the Commission decided that new ATV broadcasters would share frequency bands and channel allocations with existing analog-based TV services. The Commission also elected to adopt a "simulcast" approach, whereby the new ATV signals would be broadcast over currently unusable channels and that broadcasters would be temporarily assigned a second channel to accomplish the transition to the new ATV standard.

Six systems, four of which were all-digital, underwent extensive testing in 1991 and 1992 at the Advanced Television Test Center (ATTC) in Alexandria, Virginia. In February 1993, ACATS decided to limit further consideration to the four all-digital systems. Test results showed that all four digital systems provided impressive results, with no single digital system exhibiting a superiority that would warrant its selection over the others as the new

U.S. TV standard. ACATS decided to order supplementary tests to evaluate improvements that had been made to individual systems since initial testing.

At the same time, ACATS also adopted a resolution encouraging the competing system manufacturers to try to find a way to merge their efforts by combining the best features of each proposed digital system.

Formed in May 1993, the Digital HDTV Grand Alliance created a "best of the best" system upon which today's DTV standard is based. Members of the consortium are General Instrument Corporation, Lucent Technologies, MIT, Philips Electronics North American Corporation, the David Sarnoff Research Center, Thomson Consumer Electronics, and Zenith Electronics Corporation.

On December 24, 1996, the Commission approved the Grand Alliance's system as the new digital TV (DTV) standard for the United States (Figures 10–1 and 10–2). Since then, the DTV standard has been formally adopted by the telecommunications authorities of Canada, South Korea, and Taiwan and currently is being considered for adoption in Argentina, Australia, Brazil, China, Mexico, and Singapore.

THE DTV MODULATION SUBSYSTEM

The DTV standard employs compression technology that is based on MPEG-2 at Main Profile, High Level (MP@HL) and includes the use of Bi-directional Frame (B-Frame) motion compensation techniques that improve picture quality. The DTV digital modulation subsystem for terrestrial broadcast applications is based on 8-VSB (vestigial sideband) transmission technology, which ensures a broad coverage area, reduces interference with existing analog broadcasts, and provides immunity from interference into the digital signal. Terrestrial broadcasters will transmit at a maximum bit rate of 19.28 Mbit/s, which can support a single HDTV program or as many as five standard-definition television (SDTV) programs with a visual quality that is superior to analog NTSC TV signals.

For cable TV distribution, DTV signals can be transmitted at a higher data rate mode (38.56 Mbit/s) that uses 16-VSB to permit the transmission of two HDTV signals or multiple SDTV signals over a single 6-MHz-wide cable TV channel. The higher bit rate is possible because the cable environment is more robust than the terrestrial TV environment.

Digital DTH service providers DirecTV and USSB have announced plans to begin transmitting HDTV program services, including the new HBO HDTV channel, to U.S. subscribers via the Galaxy satellite located at 95 degrees west longitude. Satellite delivery of HDTV signals will use QPSK modulation schemes that are more suitable for transmission within broadband satellite transponder bandwidths.

Horizontal Pixels	Vertical Lines	Aspect Ratio		Picture Rate (Fields/sec)			
640	480	4:3	4:3	60 I	60 P	30 P	24 P
704	480	16:9	4:3	60 I	60 P	30 P	24 P
1,280	720	16:9			60 P	30 P	24 P
1,920	1,080	16:9		60 I		30 P	24 P

Figure 10–1 Display formats of the U.S. Digital TV (DTV) standard.

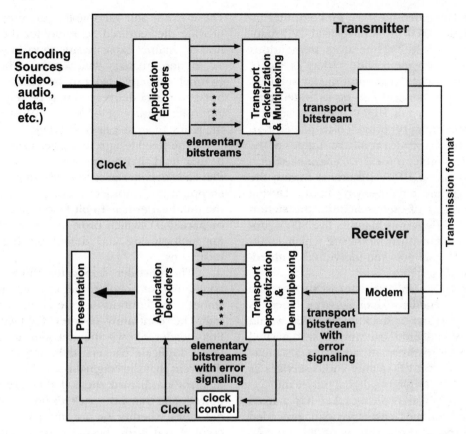

Figure 10–2 *Sample organization of functionality in a transmitter–receiver pair for a single Grand Alliance HDTV program.*

THE DTV COMPRESSION SYSTEM

Compression is an essential component of any high-definition TV transmission system. If we assume the use of 8 bits for the video luminance component and 4 bits for each of the two color difference components (Cr and Cb), we can see that the transmission of 60 progressively scanned pictures per second would require an uncompressed data rate of almost 2 Gbits/s for the active video only:

1,080 lines × 1,920 pixels × 60 frames per s ×16 bits (8 luminance and 8 chrominance) = 1,990 Mbits/s

It readily can be seen that a compression ratio in the order of 50:1 is required to transmit an HDTV signal within the 6-MHz-wide bandwidth of a single terrestrial or cable TV channel.

THE DTV TRANSPORT PACKET

Like MPEG-2, the DTV standard features a packetized data transport structure that permits the transmission of virtually any combination of video, audio, and data packets, as well as the optional selection of progressive rather than interlace video scanning for computer interoperability. This gives terrestrial TV broadcasters and cable TV operators tremendous flexibility to provide a wide variety of video, audio, voice, data, and multimedia services. Many of these services can be provided concurrently with a full HDTV program service, while others may be provided in place of an HDTV program service at different times of the day. For example, a local TV channel station could broadcast HDTV programs during the evening "prime time" viewing hours. During other portions of their schedule, the station may elect to deliver as many as five SDTV programs simultaneously, some of which could provide local schools and individuals with educational TV services.

The DTV standard also can support the delivery of ancillary data services, such as weather forecasts or stock quotes, that would be available only to those viewers who wished to subscribe to them. Broadcasters also may elect to transmit CD-quality audio services as part of the 19.28 Mbits/s digital bitstream.

Each DTV transport packet has a fixed length and contains a data "payload" preceded by a transport header that identifies the contents of each packet and the nature of the data that it carries. The transport header field contains both a fixed-length link layer and a variable-length adaptation layer (Figure 10–3).

These fixed- and variable-length components provide the required flexibility for the allocation of channel capacity among various video, audio, and auxiliary data services. The entire channel capacity also can be reallocated in bursts for data delivery as may be required to authorize a universe of decoders just prior to the airing of a pay-per-view event.

The fixed-length link layer uses a 4-byte header field that begins with the "sync_byte" that each decoder uses to establish packet synchronization. Another important element in the link header is a 13-bit field called the PID or packet ID, which provides the mechanism for multiplexing and demultiplexing digital bitstreams.

The encoder duplicates those packets containing information that is essential to the smooth and continuous operation of the system. The "continuity_counter" field within the link header allows the decoder to identify these duplicate packets, then access them in the event that the original packet was received in error or discard them if they are not required. The transport packet's link header also describes whether or not any payload is encrypted. If it is, the header flags the algorithm or electronic "key" that the decoder must use to unscramble the payload contents.

The adaptation layer is a variable-length field that handles the synchronization of the

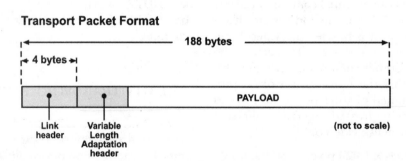

Transport Packet Format

Figure 10–3 The transport bitstream consists of fixed-length packets with a fixed and a variable component to the header field.

real-time decoding and presentation processes for each program contained within the digital bitstream. The adaptation headers of selected packets transmit the timing information that the decoder requires to maintain presentation/display synchronization. The "program clock_reference" (PCR) field contains a sample of a 27-MHz clock, which indicates the expected time at which the decoder completes the reading of that field. The decoder compares the phase of its internal or "local" clock to the PCR value to determine whether or not the decoding process is synchronized. The PCR value therefore serves as a reference that the decoder uses to adjust the clock rate.

The adaptation layer also identifies fixed points in the elementary bitstream at which insertion of local programming (e.g., commercials) is allowed, includes capabilities for supporting new functionality, and defines data that is private and with a format and meaning that is not defined in the public domain.

DTV SCANNING FORMATS

The DTV standard does not require broadcasters to use a specific scanning format, aspect ratio, or number of lines of video resolution. Instead, DTV offers each broadcaster a variety of options from which to choose. The available video formats include 24-, 30-, and 60-frame-per-second progressive scan with an advanced-definition 1,280 × 720 (number of active picture elements per line times the number of active lines) matrix of samples, and 24- and 30-frame-per-second progressive scan with a high-definition 1,920 × 1,280 matrix of samples. The DTV system also is capable of generating a 60-frame-per-second interlaced scan with a 1,920 × 1,080 matrix of samples to allow for the migration to a 60-frame-per-second 1,920 × 1,080 progressive scan format as soon as it becomes technically feasible to do so. The 60- and 30-frames-per-second rates best accommodate video source material using interlace scanning; the rate of 24 frames per second is advantageous for the transmission of all film-based source materials.

Standard-definition 640 × 480 and 704 × 480 sample matrices also are available. Although the NTSC standard is a 525-line system with 756 pixels per line, only 483 of these lines are "active" lines, with the remaining "inactive" lines contained in the vertical blanking interval.

The DTV standard features square pixels and progressive scanning so that the new DTV sets can readily interact with various personal computer platforms. Companies such as Zenith Electronics and Intel Corporation are developing PCI demodulator cards that will allow personal computers (PCs) to receive DTV broadcasts.

THE DTV AUDIO STANDARD

One major difference between an MPEG-2 DVB-compliant signal and a DTV signal is that the former uses a modified version of MUSI-CAM for the creation of CD-quality digital audio, whereas DTV uses the 5.1-channel Dolby AC-3 surround sound compression system that state-of-the-art theater sound systems already use throughout the world.

Dolby AC-3 samples the audio at 48 kHz, which is locked to the DTV 27-MHz master clock system, with a bit rate maximum of 384 Kbits/s. Dolby AC-3 supports a wide variety of primary and ancillary audio services. The five-channel complete main (CM) service features left, center, right, surround sound left, and surround sound right audio channels. A low-frequency enhancement (LFE) also is available that has a frequency response range of 3 Hz to 120 Hz. A music and special effects (ME) service also is available that supports a separate dialogue channel that can be used to deliver a secondary-language audio sound tracks. Other audio services that are supported by Dolby

AC-3 include commentary, emergency broadcasting, voice-over, and auxiliary channels for the visually and hearing impaired.

KEY TECHNICAL TERMS

The following key technical terms were presented in this chapter. If you do not know the meaning of any term presented below, refer back to the place in this chapter where it was first presented or refer to the Glossary before taking the quick check exercises that appear below.

> advanced-definition television (ADTV)
>
> Advanced Television Standards Committee (ATSC)
>
> aspect ratio
>
> digital terrestrial television broadcasting (DTTB)
>
> Dolby AC-3
>
> DTV standard
>
> DTV transport packet
>
> fixed-length link layer
>
> high-definition common image format (HD-CIF)
>
> high-definition television (HDTV)
>
> matrix of samples
>
> program clock reference (PCR)
>
> progressive scan
>
> standard-definition television (SDTV)
>
> sync byte
>
> variable-length adaptation layer

QUICK CHECK EXERCISES

Check your comprehension of the contents of this chapter by answering the following questions and comparing your answers to the self-study examination key that appears in the Appendix.

Part I: True or False

Mark each statement below "T" (true) or "F" (false) in the blank provided.

_____ 1. DTV will offer broadcasters the option of delivering their signals in either the standard 5:4 aspect ratio or in a wide-screen, 12:9 aspect ratio that more faithfully reproduces the dimensions of film-based materials.

_____ 2. The U.S. DTV standard supports various arrays of horizontal lines and vertical picture elements or "pixels" that can be displayed on the TV screen.

_____ 3. The ITU's HD-CIF format is characterized by the use of a single matrix of samples (1,920 pixels by 1,080 lines) irrespective of field and frame rate.

_____ 4. The MPEG-2 High Level and High-1440 Level can both support high-definition (HDTV) and advanced-definition TV (ADTV) pictures with $1,920 \times 1,080$ and 960×576 sample matrices.

_____ 5. A single MPEG-2 transport stream is not capable of delivering standard TV and HDTV signals simultaneously.

_____ 6. The transition from analog to digital TV for terrestrial delivery of broadcast signals means that more than 1.2 billion TV sets worldwide eventually will need to be replaced or augmented with digital set-top converter boxes.

Quick Check Exercises
Answer Key

CHAPTER 1

Part I: Matching Questions

1. The spot on the Earth's equator over which a geostationary satellite is positioned is called the subsatellite point.

2. One cycle per second is also called a hertz.

3. A frequency of 1,000 cycles per second is also called a kilohertz.

4. A combination of an uplink receiver and a downlink transmitter is called a satellite transponder.

5. Ku-band satellites transmit within the 10.7–12.75 GHz frequency spectrum.

6. A frequency of 1,000,000 cycles per second is also known as a megahertz.

7. The effective isotropic radiated power (EIRP) of a satellite signal is expressed in decibels referenced to 1 watt of power.

8. S-band satellites transmit using frequencies in the 2.6-GHz frequency band.

9. The beam width of a satellite antenna is a function of signal wavelength and antenna diameter.

10. The wavelength of a communications signal can be determined by dividing the speed of light by the signal's frequency.

Part II: True or False

11. The apparent spacing in degrees between two adjacent geostationary satellites is always smaller than the

actual spacing between the two spacecraft in degrees of longitude. False. The apparent spacing is always more than actual spacing in degrees of longitude.

12. "Extended" C-band transponders operate within the 3.45–3.7 GHz and 4.2–4.8 GHz frequency ranges. True.

13. As frequency increases, antenna beam width decreases. True.

14. As frequency increases, wavelength decreases. True.

15. The beam width of a 1.2-m antenna is narrower when receiving S-band signals than it is when receiving Ku-band signals. False.

16. The apparent spacing between two geostationary satellites is always less than the spacing in degrees of longitude from one orbital location to the next. False.

17. The super high frequency bands are located between 2.5 GHz and 22 GHz. True.

18. Satellite transponder bandwidths typically range from 24 kHz to 108 kHz. False.

19. Most geostationary communication satellites transmit signals using two orthogonal senses of polarization so that each satellite can reuse the available satellite spectrum twice. True.

Part III: Multiple Choice

20. Signals propagated within the 3–30 MHz frequency range are also known as short waves.

21. Communication satellites dowlink their signals within the C- and Ku frequency bands.

22. The narrow corridor within which a satellite antenna looks up at the sky is also known as the antenna beam width.

23. The beam coverage pattern that can cover 42.4 percent of the Earth's surface from any given location in geostationary orbit is called the global beam.

24. The signal level that would be generated if the satellite transponder is operated at full power is called saturation.

25. The satellite spectrum that typically is used for the high-power transmission of direct-to-home TV signals is called the Ku-band.

CHAPTER 2

Part I: Matching Questions

1. Digital video compression systems using B frames achieve a higher level of bit rate efficiency but require that the decoder possess a second buffer memory circuit, which adds to the cost of the receiving system.

2. A single frame of PAL video contains 625 lines.

3. Digital DTH satellites typically transmit digitally compressed video at bit rates of 1.5–8 megabits per second (Mbit/s).

4. Digital satellite transmission systems usually use some form of forward error correction (FEC) in order to improve the accuracy of the received data.

5. The "threshold" level of a digital satellite TV receiver is usually specified as a bit error rate (BER).

6. The MPEG-1 digital video compression standard is used to process all progressive scanning sources of media, such as film-based materials, text, and computer graphics.

7. Two essential components of any video signal are the black-and-white or luminance information as well as the color or chrominance information.

8. Motion compensation is used to compute the direction and speed of moving objects in a digitally compressed video image.

9. Digital satellite transmissions using MCPC combine numerous video, audio, and data signals into a single digital bitstream or multiplex.

10. The small difference between each predictor macroblock and the affected current macroblock is called the motion-compensated residual.

Part II: True or False

11. The MPEG-2 standard was developed to handle the digital compression of all media using interlaced scanning techniques, such as broadcast TV signals. True.

12. Convolutional and block encoding techniques are used by several Asian broadcasters to prevent unauthorized access to satellite TV signals that are transmitted in an analog format. False.

13. MPEG compression relies on preprocessing techniques to remove those components of a video image that are not essential to human perception. True.

14. The ultimate goal of every digital DTH installer is to fine-tune the antenna and feed in order to achieve as low a receiver bit error rate (BER) as possible. For example, a BER of 3×10^{-3} (3 E–3) would be vastly superior to a BER of 6×10^{-7} (6 E–7). False.

15. Video resolution quality is determined by the number of active video lines being transmitted as well as the number of picture elements or "pixels" contained within each line. True.

16. All satellite TV service providers who use digital video compression must multiplex all of their video, audio, and data signals into a single digital bitstream before uplinking to any given satellite. False. Digital SCPC transmissions are not required to do this.

17. Variable-length coding often is compared to the Morse code system because frequently transmitted messages are assigned the longest code sequences while infrequently transmitted messages are assigned very short code sequences. False.

CHAPTER 3

No exercises provided for this chapter.

CHAPTER 4

Part I: True or False

1. The higher the noise temperature of the C-band LNB, the better its performance. False.

2. The gain of the LNB output is the ultimate figure of merit for calculating its performance. False. It's the noise temperature (C-band) or noise figure (Ku-band).

3. The illumination taper of the feedhorn controls the signal contribution from various parts of the antenna's reflector. True.

4. The focal length is the distance between the lip of the feedhorn and the antenna's rim. False. It's the distance between the center of the reflector and the throat of the feedhorn.

5. Offset-fed antennas position the feedhorn out of the path of the incoming signal. True.

Part II: Matching Questions

6. The focal length is the measurement from the focal point to the center of the dish.

7. The gain of an LNB is usually 50–60 dB.

8. Ku-band LNBs are rated according to their noise figure while C-band LNBs are rated according to the noise temperature.

9. The ratio of focal length to antenna diameter is called the f/D ratio.

10. Sidelobes of lower intensity are part of every parabolic antenna's radiation pattern.

Part III: Multiple Choice

11. You have just installed a C-band system to receive a signal with an EIRP of 31 dBW. The best combination of antenna diameter and LNB noise temperature is 3 m/30 K.

12. You have just installed a Ku-band system to receive a signal with an EIRP of 42 dBW. The best combination of antenna diameter and LNB noise figure is 1.2 m/1.2 dB.

CHAPTER 5

Part I: Matching Questions

1. A correction factor called the declination must be incorporated into the modified mount to tilt the antenna downward toward the geostationary arc.

2. The gain of an antenna is an expression of the signal amplification provided by the reflector.

3. The focal length is the measurement from the focal point to the center of the antenna reflector.

4. The figure of merit for any satellite receiving system is its C/N rating expressed in dB.

5. A parabolic antenna has the ability to reflect all incident rays arriving along the antenna's axis of symmetry to a common location called the focal point that is located at the front and center of the dish.

6. The beam width of an antenna determines the area of the sky that is received by the antenna's main beam, while the sidelobes are secondary beams of lower intensity that are capable of picking up signals from satellites adjacent to the selected one.

Part II: True or False

7. The gain of a parabolic antenna is the most important figure of merit for evaluating its performance. False. It's the G/T.

8. Antenna noise temperature is a function of the elevation angle of the dish. True.

9. The primary advantage of a cassegrain antenna is that its subreflector blocks a lower percentage of the antenna's total surface area than the feedhorn blocks on a prime focus antenna. False. The primary advantage is that the cassegrain's feed is mounted in the center of the dish and therefore looks up toward the cold sky rather than down at the hot Earth.

10. The G/T for any satellite TV receiving system is determined by dividing the

antenna gain in decibels by the noise temperatures of the LNB and antenna. False. It's determined by subtracting the system (LNB and antenna) noise in decibels from the antenna gain in decibels.

CHAPTER 6

Part I: True or False

1. The conditional access component of an IRD is contained in each unit's smart card. True.

2. So-called "hard" encryption systems remove the horizontal and vertical sync pulses and invert the video to prevent unauthorized reception. False.

3. Adjustable bandwidth filters are an essential feature of any digital IRD. False.

4. The threshold of an analog IRD is the point at which the relationship between the C/N of the incoming satellite signal and the S/N of the displayed video departs from a linear relationship. True.

5. Encryption of digital satellite TV signals takes place at the block level. True.

6. DVB-compliant digital IRDs are mutually interchangeable. False.

7. The "seed" element of an encryption system is transmitted over the satellite along with the program signal. True.

8. The IRD's pseudorandom binary sequence (PRBS) generator controls the implementation points for video encryption. True.

9. The digital IRD must be capable of tuning to the signal's exact transmission rate in megasymbols. True.

10. The threshold of a digital IRD is the point at which the relationship between bit error rate and MPEG-2 data rate is no longer linear. False. Digital threshold is defined as a BER.

Part II: Multiple Choice

11. Analog encryption system that cuts video lines and then repositions the individual segments: (d) line translation.

12. The numerical key that the encoder can frequently change at will is called the: (a) control algorithm.

13. Each bouquet's digital signals consist of a unified bitstream, or: (b) multiplex.

CHAPTER 7

Part I: True or False

1. The inclination angle of the feedhorn opening and the inclination angle of the rim of the dish should be the same. True.

2. The distance from the rim to the dish to the feedhorn's waveguide opening from various points along the rim should be equal. True.

3. The angle of the modified polar mount's polar axis and the rim of the dish should be the same, regardless of site location. False.

4. The focal length is shorter for a so-called deep dish than it would be for a shallow dish of the same diameter. True.

5. The adjustment of the feedhorn's scalar ring plate controls the polarization alignment of the receiving system. False. It matches the feed to the f/D ratio of the antenna design.

6. The adjustment of the feedhorn's skew sets the polarization alignment to the receiving system. True.

7. The modified polar mount's declination offset angle is a function of the physical latitude of the site location. True.

8. The mount's polar axis angle is a function of the physical longitude of the site location. False.

9. An antenna with a diameter of 3 m and a focal length of 120 cm would have an *f/D* radio of 0.4. True.

10. Knowledge of the site's magnetic correction factor is essential in determining the actual inclination angle of the dish. False.

11. The mount's polar axis should be aligned to true north for locations south of the equator or true south for locations north of the equator. True.

12. The azimuth angle for any satellite visible from the site location can be determined by using a compass and the inclination/elevation angle determined by using a level or plumb bob. False.

13. To prevent moisture from entering the LNB's "F" connector, electrical tape should be wrapped tightly around the connector and nearby area of coaxial cable. False.

14. During the site survey, all compass readings should be taken out in the open, away from overhead electrical lines and high-power transformers. True.

15. Site locations on the Earth's equator will not require *any* declination offset angle. True.

16. Line amplifiers are recommended to boost signal strength over long lengths of cable because they are cheaper than using a lower loss coax, which is very expensive. False. It might be cheaper, but it is not recommended.

Part II: Multiple Choice

17. Antenna performance can be peaked by taking a signal reading with a: (a) spectrum analyzer and (b) signal turning meter.

18. The antenna mount angle that tilts the antenna downward slightly towards the Earth's equator is called the (d) declination angle.

19. The antenna mount angle that is a function of site latitude is called the (d) declination angle.

20. To determine the correct azimuth bearing when using a compass, the magnetic factor for the site location should be (c) either added to or subtracted from the compass reading.

21. The arc zenith angle at the site location is (a) the antenna declination offset angle plus the polar axis angle.

CHAPTER 8

Part I: Matching Questions

1. The ATM architecture uses a low-speed modem connection to initiate subscriber requests and a high-speed satellite channel to deliver the requested information to the subscriber.

2. The pull mode of delivery requires that each subscriber have a local ISP connection for the initiation of all subscriber information requests.

3. TCP/IP is a suite of protocols that allows communication between groups of dissimilar computer platforms.

4. The datagram is a data packet that carries information from its source to the destination over the Internet.

5. The MPEG-2 DSM-CC specification serves as the heart of the DVB data broadcasting specification.

6. A modem is the communications device that allows personal computers to connect to the Internet over traditional telephone lines.

7. The one-way satellite broadcast of multimedia content to subscribers is known as push mode.

8. The two-way interactive broadcast of multimedia content to subscribers is known as pull mode.

CHAPTER 9

Part I: Matching Questions

1. The processed IF distribution method uses a stacked LNF to produce a wideband IF signal containing signals of both senses of polarization. This signal is sent to each dwelling's IRD via a single coaxial cable.

2. The SMATV head end is the central processing center for all signals that are received by the system's terrestrial and satellite antennas.

3. QAM is a form of amplitude shift keying where the amplitude and the phase of a series of baseband pulses are modulated to represent the message.

4. The TDT is the system component that converts the incoming satellite signals from a QPSK format to a quadrature amplitude modulation (QAM) signal for distribution to individual residences connected to the SMATV system.

5. The frequency band containing information, either prior to the modulation of the information onto a radio-frequency (RF) carrier or following

demodulation of the signal at the IRD, is called the baseband.

6. The "32" in 32-QAM refers to the number of discrete signal-state values of vector magnitude that the QAM-modulated signal supports.

Part II: True or False

7. The purpose of the transparent digital transmodulator (TDT) is to convert QAM signals into a QPSK modulation format. False. It's to convert QPSK signals into a QAM modulation format.

8. DVB-CS is a set of specifications pertaining to SMATV systems that was established by the Digital Video Broadcasting (DVB) Group. True.

9. QAM is a more effective modulation scheme than QPSK for the distribution of digital signals in a bandwidth-limited environment such as SMATV, where the bandwidth of a single channel is limited to just a few megahertz. True.

10. Digital transmodulation is the process whereby signals of opposite polarization from a given satellite are multiplexed onto a single coaxial cable through the use of a transparent digital transmodulator at the SMATV head end. False.

Part III: Multiple Choice

11. The Digital Video Broadcasting (DVB) Group has established the following specifications governing the use of the MPEG-2 digital compression standard for SMATV distribution purposes: (a) DVB-CS.

12. An all-digital SMATV head end may contain the following components: (a) a and b—transparent digital transmodulator (TDT) and an IRD.

13. The SMATV system normally uses a larger satellite antenna than what would be required at for a home satellite TV system because: (e) b and c—the SMATV system must generate a very good signal at the system head end to reduce attenuation and other degradations in the cable distribution system and the larger dish will give the system a greater signal margin to counteract rain fades.

14. The "16" in 16-QAM refers to: (d) the number of discrete signal-state values of vector magnitude that the QAM signal supports.

15. The transparent digital transmodulator converts the satellite's QPSK-modulated digital signal into an equivalent: (b) QAM signal that can be distributed through a 6-MHz wide (North America) or 8-MHz wide (Europe) cable TV channel.

CHAPTER 10

Part I: True or False

1. DTV will offer broadcasters the option of delivering their signals in either the standard 5:4 aspect ratio or in a wide-screen, 12:9 aspect ratio that more faithfully reproduces the dimensions of film-based materials. False.

2. The U.S. DTV standard supports various arrays of horizontal lines and vertical picture elements or "pixels" that can be displayed on the TV screen. True.

3. The ITU's HD-CIF format is characterized by the use of a single matrix of samples (1,920 pixels by 1,080 lines) irrespective of field and frame rate. True.

4. The MPEG-2 High Level and High-1440 Level can both support high-definition (HDTV) and advanced definition TV (ADTV) pictures with $1,920 \times 1,080$ and 960×576 pixel arrays. True.

5. A single MPEG-2 transport stream is not capable of delivering standard TV and HDTV signals simultaneously. False.

6. The transition from analog to digital TV for terrestrial delivery of broadcast signals means that more than 1.2 billion TV sets worldwide eventually will need to be replaced or augmented with digital set-top converter boxes. True.

Glossary of Technical Terms

Actuator. The motor that rotates or actuates the modified polar mount so that the parabolic reflector's main beam sweeps across the portion of the geostationary arc that is visible from the site location.

ADPCM (adaptive differential pulse code modulation). A compression technique that encodes the predictive residual signal instead of the original waveform. This improves the compression efficiency by only transmitting the small difference between the estimate of the next sample and the actual sample, which can be encoded in fewer bits than the actual sample.

ADTV (advanced-definition television). A wide-screen television signal with video resolution that is substantially better than what traditional TV systems deliver.

Algorithm. A mathematical process used by encryption and/or compression systems to encode video, audio, and data signals.

AM. Amplitude modulation.

Analog. Communication signals of varying frequency and/or amplitude.

Aperture. The effective capture area of a communications antenna.

Arc zenith. The highest point in the geostationary arc located along the true north/south line that crosses through the site location.

Artifacts. Visual impairments to a video signal that are the result of limitations of the video transmission standard.

ASCII. American Standard Code for Information Exchange.

Aspect ratio. The picture width to picture height ratio of the TV screen's display area.

ATSC. Advanced Television Standards Committee (USA).

Attenuation. Loss of signal through a transmission medium or component expressed in decibels.

Asynchronous transfer mode (ATM). The switched transmission of data in small, fixed-size cells that lend themselves to the time-division-multiplexing characteristics of transmission media and the packet switching characteristics of data networks. At each switching node, the ATM header identifies a virtual path or circuit for which the cell contains data so that the switch can forward the cell to the correct next-hop trunk. Two endpoints that desire to communicate set up the virtual path through the involved switching nodes.

ATV. Advanced TV. See also ADTV.

Axis of symmetry. The bisecting line that divides the paraboloid into two symmetrical halves.

Azimuth. The magnetically corrected compass bearing (360 degrees) that is used to locate a geostationary telecommunications satellite.

B frame. The bidirectional video frame in the MPEG-2 compression system, which has the ability to access and employ motion-compensated prediction from past and/or future I and P reference frames.

Band. A subsegment of the electromagnetic spectrum that national and international telecommunications authorities assign to one or more types of communications services.

Bandwidth. The subsegment of the electromagnetic spectrum in kilohertz or megahertz used to transmit one or more communications signals.

Baseband. A frequency band containing information, either prior to the modulation of the information onto a radio carrier, or following demodulation on the receive end.

Beam width. The width in degrees of the narrow corridor along which the antenna transmits and/or receives communications signals. For parabolic satellite antennas, the beam width is defined at the −3 dB (half power) points.

BER (bit error rate). The figure of merit for digital transmission systems that expresses the probability of one or more bit errors occurring in a given number of bits.

Bit. Binary digit, either 0 or 1.

Bit rate. The rate at which bits of information are transmitted in a digital communications system.

Block. An 8-row by 8-column matrix of picture elements (pixels) or the equivalent of 64 DCT coefficients. Also a predetermined group or string of binary digits.

Block coding. A digital coding system in which the encoder looks only at the bits contained within each individual block of data.

Block downconversion. The use within an LNB of a fixed frequency local oscillator (LO) that heterodynes with the incoming satellite frequency band to produce a lower intermediate frequency (IF) band for subsequent tuning and demodulation at the receiver.

Bouquet. A unified group of digitally compressed telecommunication services.

BSS. Broadcast Satellite Service.

C-band. The 3.4–6.5 GHz frequency spectrum.

Carrier. The primary frequency of a communications signal that is modulated by the baseband information.

Cassegrain. A parabolic antenna that employs dual-reflector geometry to reflect the incoming signal back to a feedhorn mounted inside the center of the main reflector.

CCIR. International Radio Consultative Committee.

CCITT. The committee of the International Telecommunication Union that is responsible for making technical recommendations about telephone and data communication systems.

Celestial equator. The projection of the Earth's equator onto the sky.

Circular polarization. Rotation of the signal wave front in either a clockwise (right-hand) or counterclockwise (left-hand) direction as seen from the signal source.

Coaxial cable. A shielded cable with an insulated center conductor that is used for passing radio-frequency (RF) signals.

Chroma format. The number of chrominance blocks in a macroblock.

Chrominance. The color component of a video signal. In digital systems this is a matrix, block, or single picture element that represents one of the two color difference signals related to the primary colors. The symbols Cr and Cb are used to signify the color difference signals.

CIF (Common Image Format). An international standard governing the structure of the samples that represent the visual information contained within a single frame of digital HDTV that is totally independent of the frame rate and the sync/blanking structure of the video signal. The uncompressed bit rate for transmitting CIF at 29.97 frames/s is 36.45 Mbit/s.

Coded order. The order in which frames of video are stored and decoded, which may not necessarily be the same as the display order.

Colocation. The location of multiple geostationary satellites at a single orbital assignment over the Earth's equator.

Composite baseband video. The unprocessed demodulator output of the analog receiver or IRD prior to filtering, clamping, and deemphasis.

Compression. Removal of nonessential and/or redundant information from a communications signal in order to reduce the amount of transmission bandwidth required. The information removed either is nonessential or can be restored at the receive end.

Concatenation. The use of two coding systems in which the output of the inner encoder falls or cascades into the outer encoder.

Conditional access (CA). The authorization data that allows a decoder to access an encrypted communications signal.

Convolutional coding. A digital coding system that incorporates memory so that the encoder can look at both previous and current blocks of data.

CRT. Cathode ray tube.

Cycle. A sine wave with 360 degrees of revolution.

Cycles per second. The measurement unit of frequency, also called a hertz after the nineteenth-century discoverer of "Hertzian waves." 1,000 cycles per second is a kilohertz; 1 million cycles per second is a megahertz; 1 billion cycles per second is a gigahertz.

C/N. Carrier-to-noise ratio of a communications signal, measured at RF or IF.

DBS. Direct broadcast satellite.

Decibel (dB). A unit of measurement that expresses changes in signal levels along a logarithmic scale: 3 dB represents a multiplication factor of 2; 10 dB a factor of 10; 20 dB a factor of 100; 30 dB a factor of 1,000; etc.

dBi. The expression of antenna gain in reference to an isotropic reference antenna.

dBm. dB milliwatt.

dBW. The expression of satellite signal strength in decibels relative to one watt of power.

DCT (discrete cosine transform). The MPEG-2 compression system uses the DCT mathematical algorithm to convert blocks from a spatial domain to an equivalent set of DCT coefficients that may be expressed in a frequency domain.

Declination. The angular distance north or south of the celestial equator. For satellite communications systems this is the correction angle incorporated into the polar mount that tilts the reflector downward slightly toward the geostationary arc. (See Polar mount.)

Decoder. The processing unit at the receive end of a communications link. The decoder receives information that has been coded at the transmit end, either to compress the signal or to encrypt the information, and restores the information to its original state.

Display order. The order in which the decoded frames of video are displayed. Normally this is the same order in which they were presented at the input of the encoder.

Downlink. The satellite receiving system. Also refers to the frequency band that the communications satellite uses to transmit signals back to Earth.

DPCM (differential pulse code modulation). A source coding scheme developed for encoding sources with memory.

DRO. Dielectric resonant oscillator. (See also LNB.)

DTH. Direct to home.

DTV. Digital television. Refers to the digital standard adopted by the U.S. Federal Communications Commission for terrestrial service within the United States.

DVB (Digital Video Broadcast). A digital compression standard that incorporates the MPEG-2 specification as a subset. Communications systems that use the DVB standard are said to be DVB-compliant.

E_b/N_o **(energy bit to noise density ratio)**. This is equivalent to C/N for digital receiving systems.

ECM. Electronic countermeasure.

EIRP. Effective isotropic radiated power.

EPG. Electronic program guide.

Electromagnetic spectrum. The entire continuum of frequencies used to propagate communication signals, from very low frequencies to visible light and beyond.

Elevation. The angle (0 to 90 degrees) at which the antenna tilts up at the sky. Also the complement of inclination.

Encoder. The unit at the transmit end that converts information mathematically, either to improve signal quality or to encrypt the information.

Encryption. A mathematical process used to encode communications signals so that only the authorized receiving stations can access the information.

ETSI. European Telecommunication Standard Institute

Extended C-band. The 3.4–3.7 GHz and 4.2–4.8 GHz frequency ranges.

FCC. Federal Communications Commission (USA).

F connector. A coaxial connector for use with coaxial cables having a characteristic impedance of 75 ohms.

f/D. The ratio of focal length to antenna diameter.

FEC. Forward error correction.

Feedhorn. A signal capturing device that is used to illuminate the reflector of a parabolic antenna.

Field. One-half of a complete vertical scan of a TV image consisting of either the even-numbered or the odd-numbered lines.

The two fields are combined, or interlaced, to make up each frame of video.

Field period. The reciprocal of twice the video frame rate.

Filter. A communications device used to limit the frequency bandwidth of a communications signal.

FM. Frequency modulation.

Focal length. The distance between the center of the paraboloid and its focus.

Footprint. The signal coverage pattern generated by a communications satellite's transmit antenna.

Frame. One complete vertical scan of a TV image. For progressive video, the lines that make up each frame contain samples starting from one time instant and continuing through successive lines to the bottom of the frame. For interlaced video, a frame consists of two alternating fields.

Frame period. The reciprocal of the frame rate.

Frame rate. The rate at which frames are output during the decoding process.

Frequency. The number of times an alternating current goes through a complete cycle of 360 degrees in one second of time. (See also Cycle and Cycles per second).

FSS. Fixed Satellite Service.

Gain. The amplification factor for communications devices expressed in decibels. For antennas, gain is expressed in dBi, decibels referenced to an isotropic reference antenna.

Gateway. The functional bridge between networks that provides for protocol translation.

Gb/s. Gigabits per second.

Geostationary orbit. A circular orbit located in the plane of the Earth's equator at a distance of approximately 22,300 miles within which satellites maintain a fixed position in the sky relative to receiving stations down on the ground.

Gigabit (Gbit or Gb). 1 billion bits.

Gigahertz (GHz). 1 billion cycles per second.

GOP. Group of pictures.

Group of pictures. A series of video frames comprising one video scene. (See also I, P, and B frame.)

G/T. The figure of merit for a receiving system, which is calculated by subtracting the system noise temperature (T) in decibels from the gain (G) in decibels produced by the receiving antenna.

HBI. Horizontal blanking interval.

HD-CIF. High Definition Common Interface Format.

HDTV. High-definition television.

Head end. The signal processing center of a cable TV or SMATV receiving system.

Header. A block of data in the coded bitstream containing the coded representation of a number of data elements pertaining to the coded data that follow the header in the bitstream.

HEMT. High electron mobility transistor.

Hertz (Hz). One cycle per second.

Huffman coding. A compression coding system that assigns short codes to frequently occurring characters and longer codes to infrequently occurring characters. This minimizes the average number of bytes required to represent the information to be transmitted.

I frame (intra frame). A frame of video in the MPEG-2 compression system that is encoded with reference only to the information contained within itself.

IEEE. Institute of Electrical and Electronics Engineers.

Illumination taper. The attenuation of the signal arriving at the feedhorn from the outer portion of the parabolic reflector.

Inclination. The complement of the antenna elevation angle (90 – elevation angle) in degrees.

Inclinometer. A device used to measure the antenna elevation angle.

Interlace. The property of conventional frames of video whereby two fields containing half the total number of video lines are alternated.

Intra coding. Coding of a macroblock or picture that uses information only from that macroblock or picture.

Ionosphere. Upper layers of the Earth's atmosphere that are electrically charged by solar radiation and therefore are capable of reflecting telecommunications signals at certain frequencies.

IRD. Integrated receiver/decoder.

ISDN (Integrated Services Digital Networks). A CCITT standard for digital telephony that uses existing switches and wiring to relay a 64 kb/s end-to-end channel.

ISO. International Standards Organization.

ITU. International Telecommunication Union.

JPEG. The digital compression standard used for computer graphics and an early precursor to the MPEG compression standard. M-JPEG (for Motion JPEG) is one derivative of the JPEG standard that was used early on to compress moving pictures.

Ka-band. The 18.3–31 GHz frequency spectrum.

Kelvin (K). Unit of measurement for thermal noise.

Kilobit (kbit or kb). 1,000 bits.

kb/s. 1,000 bits per second.

Kilohertz (kHz). 1,000 cycles per second.

Ku-band. The 10–18 GHz frequency spectrum.

L-band. The 950 MHz to 2 GHz frequency spectrum.

Letter box. A TV picture display format that limits the recording or transmission of useful picture information to about three-quarters of the vertical picture height of the distribution format (e.g., 525-line) in order to offer program material that has a wide picture aspect ratio.

Level. A defined set of constraints on the values that may be taken by some parameters within a particular profile. A profile may contain one or more levels.

LNB. Low-noise block downconverter.

LNF. Low-noise feed.

LO. Local oscillator. (See also LNB.)

Luminance. The brightness component of a video signal.

Macroblock. A 16 × 16 pixel array consisting of four blocks of 8 × 8 pixels each.

Magnetic correction factor. The difference in degrees between true north and magnetic north at the site location that is used to correct compass bearing readings.

Mb/s (or Mbps). A transmission rate of 1 million bits per second.

MCPC. Multiple channel per carrier.

MDU. Multiple dwelling unit.

Megabit (Mbit or Mb). 1 million bits.

Megahertz (MHz). A frequency of 1 million cycles per second.

Megasymbol (Msym). 1 million symbols.

Microwaves. Super high frequency (SHF) transmissions above 3 GHz.

Modem (modulator/demodulator). An electronic device that converts serial data from a computer to an audio signal that can be transmitted over standard telephone lines. The audio signal usually is composed of silence (no data) or one of two frequencies representing 0 and 1.

Modem baud rate. Modem transmission speeds, called baud rates, can range from 75 baud up to 56,000 and beyond.

Modified polar mount. See Polar mount.

Modulation. The process of attaching information to an RF carrier wave.

Motion-compensated residual. The relatively small difference between each predictor block and the current block in an MPEG-2 compression system.

Motion compensation. In MPEG compression systems, prediction of the motion vectors of macroblocks from one frame to the next.

Motion estimation. The process of estimating motion vectors during the MPEG encoding process.

MPEG. Motion Pictures Experts Group.

MPEG-1. Compression system for progressive scan sources of media such as text, graphics, and film.

MPEG-2. Compression system for interlace scanning based sources of media such as broadcast TV.

Msym/s. A transmission rate of 1 million symbols per second.

Multiplex. The combination of multiple video, audio, and data signals into a single unified digital bitstream.

Noise figure. A measurement of thermal noise expressed in decibels.

Noise temperature. A measurement of thermal noise expressed in kelvins (K).

NTSC (National Television System Committee). U.S. video standard with a 4:3 aspect ratio, 525 lines, 60-Hz frame rate, and 4-MHz video bandwidth with a total 6 MHz of video channel width.

Offset angle. The deviation in degrees from the axis of symmetry of a paraboloid reflector.

Offset-fed antenna. A derivation of a prime focus parabolic antenna that uses a subsection of the curve that locates the focus away from the center of the reflector.

Orthogonal. Used to describe the "mutually at right angles" isolation between two opposite senses of polarization—either horizontal and vertical, or right-hand and left-hand circular.

OSI (Open Systems Interconnection). A reference model of the International Standards Organization (ISO) that established an open set of international communications standards. The OSI model consists of seven architectural layers: the physical layer; the data link layer; the network layer; the transport layer; the session layer; the presentation layer, and the application layer.

P frame. The predicted frame of video in the MPEG-2 compression system that is coded using motion-compensated prediction from past reference pictures.

Packet. A fixed-length string of binary digits containing one part of a complete message, with each packet containing a header and a check sum. Packets are transmitted independently in a store-and-forward manner.

PAL (Phase Alternating Line). European video standard with a 4:3 aspect ratio, 625 lines, 50-Hz frame rate, and 4-MHz video bandwidth within a total channel bandwidth of 8 MHz.

PCM (pulse code modulation). A coding technique in which the input signal is represented by a given number of fixed-width samples per second.

PEL. Picture element.

PES. Packetized elementary stream.

Phase. The number of electrical degrees by which one wave leads another.

Phase noise. The short-term instability of an RF signal.

PID. The picture identification number used to identify the location of an individual video service within a DVB-compliant digital bitstream.

Pixel. Picture element.

Planar array. A flat satellite antenna that employs an array of resonant elements connected in phase to capture the incoming signals.

Polarization. The property of orienting the wavefront of a communications signal in a direction or sense of rotation.

Polar mount. Antenna support structure that steers the reflector through the geostationary arc by rotation about a single axis. For satellite receiving systems, the polar mount must be modified to include the declination offset angle required to accurately access satellites in geostationary orbit.

PRBS. Pseudorandom binary sequence.

Prime focus. A parabolic antenna that produces a focus directly to the front and center of the reflector.

Profile. A defined subset of the syntax of a specification.

QAM. Quadrature amplitude modulation.

QPSK. Quadrature phase shift keying.

Quantization. In MPEG digital compression systems, the conversion of DCT coefficients into a more compact representative form.

Rain fade. Loss of signal due to the absorption and depolarization effects of raindrops in the atmosphere.

Random access. The process of beginning to read and decode the coded bitstream at an arbitrary point.

RARC. Regional Administrative Radio Conference.

Reed–Solomon. An FEC coding technique used by all DVB-compliant satellite transmission systems.

RGB. The red (R), green (G), and B (blue) picture components of a color video signal.

S-video. A cable output standard for video signals that uses a 4-pin mini-plug connector that bypasses the comb filter in a device that separates the brightness or Y and color or C components.

Scalability. The ability of a decoder to decode an ordered set of bitstreams to produce a reconstructed sequence. The first bitstream in the set is called the base layer. Each of the other bitstreams in the set are called enhancement layers.

SCPC. Single channel per carrier.

SECAM (Sequence with memory). European video standard with a 4:3 aspect ratio, 625 lines, 50-Hz frame rate.

Seed. The component of an encrypted signal that is used to synchronize the encoder and decoder.

SHF. Super high frequency.

SID. Sound identification number used to identify the location of a sound service in a DVB-compliant digital bitstream.

Sidelobe. The off-axis response of a parabolic reflector.

Skew. Deviation from true horizontal or vertical polarization as perceived from the receiving site location.

Slice. A series of macroblocks.

Smart card. An electronic countermeasure (ECM) existing in a transportable physical medium that is used by various conditional access systems.

SMATV. Satellite Master Antenna TV System.

SMPTE. The Society of Motion Picture and Television Engineers.

SNR. Signal-to-noise ratio.

Sparklies. Impulse noise that appears as comet-tailed dots in the TV picture produced by analog TV receivers operating at a low value of C/N.

Standard C-band. The 3.7–4.2 GHz frequency spectrum.

Stationkeeping. The expenditure of fuel required to keep a geostationary satellite within [plus or minus]0.1 degree of its assigned orbital assignment.

Subsatellite point. The spot on the Earth's equator over which one or more geostationary satellites are located.

Sync. The synchronization pulse used to turn off the electron gun during line or field retrace.

Threshold. The figure of merit for analog satellite receivers and IRDs expressed as a C/N value in decibels. The threshold for digital receivers and IRDs is expressed as a bit error rate (BER).

TI. Terrestrial interference.

Transponder. The combination of an uplink receiver and downlink transmitter that acts as a repeater of a communications signal.

UHF. Ultra high frequency.

Uplink. A satellite transmission system. Also may refer to the frequency that the satellite uses to receive signals from Earth.

VBI. Vertical blanking interval.

VHF. Very high frequency.

Video sequence. A series of one or more pictures.

VLC (variable-length coding). A coding technique that assigns shorter code words to frequent events and longer code words to less frequent events.

WARC. World Administrative Radio Conference of the ITU.

Watt. The unit of power measurement.

Wavelength. The physical distance that a radio wave travels during one complete cycle.

X-band. The 7–8 GHz frequency spectrum.

Zigzag scanning. The sequential ordering of DCT coefficients from the lowest spatial frequency to the highest.

Index

Page references followed by "t" denote tables; "f" denote figures

About the CD-ROM...

This Companion CD features the contents of *The Digital Satellite TV Handbook* in hyperlinked html format. To access the material, you will need to have a web brower installed on your computer. If you already have a web browser installed, launch the application and open the file **D:\index.html** (where "**D**" is the designation of your CD drive.)

If you do not have a web browser installed, you can install **Microsoft Internet Explorer 5.0** directly from the CD by running **D:\IE5-BH\IE5Setup.exe** (where "**D**" is the designation of your CD drive.) Follow the directions on screen to complete the installation. After successfully installing the web browser, launch the application and open the file **D:\index.html** (where "**D**" is the designation of your CD drive.)

The Digital Satellite TV Handbook **Companion CD** also includes the **Satmaster Pro Mk5.4 Demo** and **User Guide**.

Satmaster Pro is an easy-to-use set of software tools for the satellite professional who wants to save hours, perhaps days, of tedious work. It includes full up-down digital and FM link budgets, solar outage prediction, antenna aiming, dual feed and polar mount modules plus various graphing and table generation features. Satmaster Pro is used extensively throughout the satellite industry and is often said to be more powerful and easier-to-use than packages costing up to 10 times the price.

Uses include satellite TV broadcasting, VSAT systems, SNG, radio feeds and TVRO. The calculation methods adopted have withstood the test of time and are fully global in operation. Fixed price multi-user company licenses are also available. The 30-day evaluation demo has limited town/city data and disabled file saving and printing. It requires a 486 or better CPU running Windows 9x/2000 or NT4 (Intel). See the file UserGuide.rtf for further information.

To install the demo, run **D:\Demos\Smdemo.exe** (where "**D**" is the designation of your CD drive.)

For technical assistance with this CDR, email **techsupport@bhusa.com**. Be sure to reference **CD-71718-PC**.